Harnessing Energy Inverter Wonders

Harnessing Energy Inverter Wonders

Alina Hazel

Noble Publishing

CONTENTS

INDEX　1

1　Chapter 1　3

2　Chapter 2　24

3　Chapter 3　46

4　Chapter 4　69

5　Chapter 5　94

6　Chapter 6　120

7　Chapter 7　142

8　Chapter 8　166

9　Chapter 9　187

INDEX

Chapter 1: Introduction to Energy Inverters

1.1 Define energy inverters and their significance in the modern world.

1.2 Brief historical overview of energy inverters.

1.3 The role of energy inverters in transforming and optimizing energy.

Chapter 2: Basics of Energy Conversion

2.1 Explain the fundamental principles of energy conversion.

2.2 Different types of energy sources and their conversion into electrical energy.

2.3 Introduce the concept of inverter technology and its role in energy conversion.

Chapter 3: Types of Energy Inverters

3.1 Explore various types of energy inverters (sine wave, modified sine wave, grid-tie, off-grid, etc.).

3.2 Discuss the advantages and disadvantages of each type.

3.3 Highlight real-world applications for different types of energy inverters.

Chapter 4: Green Energy and Inverters

4.1 Examine the integration of energy inverters with renewable energy sources (solar, wind, etc.).

4.2 Showcasing successful implementations of green energy systems with inverters.

4.3 Discuss the environmental impact and sustainability aspects.

Chapter 5: Smart Grids and Inverter Technology

5.1 Explore the role of energy inverters in smart grid systems.

5.2 Discuss the benefits of smart grid technology and how inverters contribute to grid stability.

5.3 Address challenges and potential solutions in implementing smart grids with energy inverters.

Chapter 6: Innovations in Inverter Design

6.1 Highlight recent advancements in inverter technology.

6.2 Explore innovations in materials, efficiency, and miniaturization.

6.3 Groundbreaking inverter designs and their impact on the industry.

Chapter 7: Energy Storage and Inverters

7.1 Discuss the synergy between energy storage systems and inverters.

7.2 Explore how inverters contribute to efficient energy storage and retrieval.

7.3 Highlight emerging technologies and trends in energy storage coupled with inverters.

Chapter 8: Challenges and Future Trends

8.1 Address current challenges in the field of energy inverters.

8.2 Discuss potential solutions and advancements on the horizon.

8.3 Explore future trends and the role of inverters in the evolving energy landscape.

Chapter 9: Harnessing Energy Inverters for a Sustainable Future

9.1 Summarize key takeaways from the book.

9.2 Emphasize the role of energy inverters in achieving a sustainable and resilient energy future.

9.3 Inspire readers to explore further innovations and applications in the field.

Chapter 1

Introduction to Energy Inverters

Energy inverters assume a crucial part in present day power frameworks, filling in as essential gadgets that convert direct current (DC) into exchanging current (AC). This change is fundamental for various applications, going from private sunlight based power frameworks to modern apparatus. The omnipresence of AC power in our day to day routines features the basic job inverters play in empowering the effective and consistent use of electrical energy.

The key guideline behind energy inverters lies in their capacity to change the progression of power, working with the mix of different energy sources into the power matrix. In this extensive investigation, we will dive into the complexities of energy inverters, analyzing their sorts, working standards, applications, and the developing scene of inverter innovation.

Sorts of Energy Inverters:
Energy inverters come in different sorts, each intended to take care of explicit requirements and applications. Extensively grouped, inverters can be sorted into three primary sorts: sine wave inverters, changed sine wave inverters, and square wave inverters.

Sine Wave Inverters:

Sine wave inverters produce a result waveform that intently impersonates the unadulterated sinusoidal waveform of utility-provided AC power. This kind of inverter is exceptionally liked for applications requesting an excellent AC power supply, like touchy gadgets, clinical hardware, and certain modern apparatus. The smooth and persistent waveform created by sine wave inverters guarantees similarity with a large number of gadgets, limiting the gamble of electrical impedance or harm.

Adjusted Sine Wave Inverters:

Changed sine wave inverters produce an air conditioner yield that approximates a ventured waveform. While this sort of inverter is more efficient than sine wave inverters, it may not be reasonable for gadgets with explicit power prerequisites. Changed sine wave inverters are generally utilized in applications where cost is a basic component and where the gadgets being fueled can endure a less refined AC waveform.

Square Wave Inverters:

Square wave inverters produce a result waveform that, as the name recommends, looks like a square waveform. This sort of inverter is the most affordable but on the other hand is the most un-normal because of its restricted applications. Square wave inverters are reasonable for driving fundamental machines and instruments however are by and large stayed away from for gadgets that require a more steady and refined power source.

Working Standards of Energy Inverters:

Understanding the functioning standards of energy inverters is critical for getting a handle on their usefulness and productivity in changing DC over completely to AC power. The activity of inverters includes a few key stages:

DC-to-AC Transformation:

At the center of an inverter's usefulness is its capacity to change over DC capacity to AC power. This cycle includes the utilization of electronic circuits, normally made out of semiconductors or other semiconductor gadgets, to switch the DC input in a way that repeats the ideal AC waveform.

Beat Width Tweak (PWM):

Beat width tweak is a regularly utilized procedure in present day inverters. It includes changing the width of the beats in the rearranged waveform to control the typical power conveyed to the heap. By changing the beat width, inverters can manage the voltage and recurrence of the air conditioner yield, guaranteeing similarity with various kinds of burdens.

Inverters consolidate control frameworks and criticism systems to keep up with steadiness and exactness in their result. These frameworks continually screen the result waveform, making continuous acclimations to make up for varieties in burden or information conditions. This guarantees a predictable and dependable AC power supply to associated gadgets.

Transformer Stages:

In certain inverters, particularly those intended for high-power applications, transformer stages might be incorporated to additionally refine the result voltage. Transformers help with changing the voltage levels and segregating the result from the information, giving extra assurance and upgrading the general presentation of the inverter.

Uses of Energy Inverters:

Energy inverters track down applications in a different scope of ventures and settings, adding to the proficient usage of electrical power. A few outstanding applications include:

Sustainable power Frameworks:

Inverters assume a fundamental part in sustainable power frameworks, for example, sun oriented and wind power. Sun powered chargers create DC power, which is then changed over completely to AC by inverters for combination into the network or for direct use in homes and organizations. Likewise, wind turbines produce variable DC result, and inverters guarantee a steady AC supply.

Private Power Frameworks:

In private settings, inverters are regularly utilized related to sunlight based chargers to change over the DC power produced by the boards into AC power for family utilization. This permits mortgage holders to outfit sunlight based energy and decrease their reliance on conventional framework power, adding to manageability endeavors.

Uninterruptible Power Supplies (UPS):

Uninterruptible Power Supplies depend on inverters to give a consistent progress from matrix capacity to battery power during blackouts. Without a trace of matrix power, the inverter changes over the DC power put away in batteries into AC power, guaranteeing a consistent and continuous power supply to basic gadgets and frameworks.

Modern Applications:

Enterprises utilize inverters for different applications, including the activity of engine drives, control frameworks, and apparatus. Inverters consider exact control of engine speed and force, adding to energy proficiency and cycle advancement in modern cycles.

Electric Vehicles (EVs):

Energy inverters are basic parts in electric vehicles, changing over the DC power put away in the vehicle's batteries into AC ability to drive the electric engine. The productivity and execution of the inverter essentially influence the general effectiveness and scope of electric vehicles.

Media communications:

In the field of media communications, inverters are utilized in reinforcement power frameworks. During blackouts, inverters guarantee a consistent progress to battery power, empowering constant activity of basic correspondence gear.

Crisis and Remote Power:

Inverters are fundamental in crisis and remote power situations where admittance to lattice power might be restricted. Versatile inverters can change over DC power from batteries, generators, or sunlight based chargers into AC power for different applications, including setting up camp, crisis lighting, and far off building locales.

Development of Inverter Innovation:

The field of inverter innovation has seen huge headways through-out the long term, driven by the developing interest for effective and dependable power change. A few critical patterns and improvements in inverter innovation include:

Further developed Productivity:

Progressing innovative work endeavors have zeroed in on improving the effectiveness of energy inverters. Higher productivity not just decreases energy misfortunes during the change cycle yet additionally adds to the general maintainability of influence frameworks.

Scaling down and Incorporation:

Progressions in semiconductor innovation have prompted the scaling down and reconciliation of inverter parts. This pattern brings about additional conservative and lightweight inverters, making them reasonable for a more extensive scope of uses, including convenient gadgets and electric vehicles.

Savvy Inverters:

The joining of brilliant advancements, like microcontrollers and computerized signal processors, has empowered the improvement of savvy inverters. These inverters can speak with different gadgets, adjust to changing burden conditions, and give constant information on energy creation and utilization.

Matrix Intelligent Inverters:

With the rising mix of sustainable power sources into the frame-work, network intelligent inverters have acquired unmistakable quality. These inverters can synchronize with the framework, taking care of overabundance energy back into the network and supporting lattice soundness.

High level Control Calculations:

Current inverters utilize complex control calculations to improve execution and address dynamic working circumstances. These calcula-tions empower fast reaction to changes in load, guaranteeing a steady and solid power supply.

Improved Dependability and Strength:

Continuous exploration in materials science and designing has prompted the advancement of inverters with further developed unwavering quality and sturdiness. This is especially significant in applications where inverters are exposed to cruel ecological circumstances or regular cycling.

Staggered Inverters:

Staggered inverters have become progressively well known in high-power applications. These inverters utilize different voltage levels to create a flight of stairs like waveform, lessening symphonious mutilation and working on the nature of the result waveform.

Joining with Energy Stockpiling:

Inverters are presently much of the time incorporated with energy capacity frameworks, like batteries. This reconciliation takes into account better administration of energy stream, empowering the capacity of abundance energy during times of low interest and its delivery during top interest.

Difficulties and Future Possibilities:

While energy inverters have taken exceptional steps as far as proficiency and usefulness, a few difficulties and valuable open doors lie ahead:

Matrix Coordination Difficulties:

As the portion of sustainable power sources in the power framework keeps on developing, challenges connected with network mix might emerge. Savvy inverters furnished with lattice steady elements are pivotal for keeping up with matrix strength and unwavering quality.

Energy Capacity Mix:

Incorporating inverters with energy capacity frameworks presents both specialized and financial difficulties. Creating savvy and proficient answers for putting away and recovering energy stays a center region for future examination.

Normalization and Interoperability:

Normalization of inverter innovations and guaranteeing interoperability among various frameworks are fundamental for making a consistent and interconnected energy scene. Broad guidelines can work with the incorporation of different energy sources and capacity frameworks.

Online protection Concerns:

With the rising dependence on computerized advances, network protection is a developing worry in the energy area. Inverters, being basic parts in power frameworks, need vigorous network safety measures to safeguard against expected dangers and weaknesses.

Progressions in Semiconductor Materials:

Proceeded with progressions in semiconductor materials, for example, wide-bandgap semiconductors, hold the possibility to additionally work on the effectiveness and execution of energy inverters. These materials offer benefits like higher temperature resilience and diminished energy misfortunes.

Half breed Energy Frameworks:

The reconciliation of different energy sources and capacity advances in crossover energy frameworks presents energizing open doors for inverter innovation. Cross breed frameworks that join sun based, wind, and other inexhaustible sources with conventional power sources require flexible and versatile inverters.

Worldwide Zap Drives:

With regards to worldwide zap drives, energy inverters assume a pivotal part in stretching out admittance to power in remote and underserved districts. Creating financially savvy and solid inverters customized to the particular necessities of these districts is really important.

Inverter Reusing and Maintainability:

As the quantity of inverters being used increments, tending to end-of-life issues and reusing contemplations becomes foremost. Reasonable practices in assembling, reusing, and removal of inverters add to limiting natural effect.

1.1 Define energy inverters and their significance in the modern world.

Energy inverters address crucial parts inside contemporary power frameworks, filling in as fundamental gadgets that work with the change of direct current (DC) into substituting current (AC). This extraordinary interaction is of fundamental significance across different applications, traversing private sunlight based power establishments to mind boggling modern hardware. The universality of AC power in our day to day routines highlights the basic job that inverters play in empowering the productive and consistent use of electrical energy.

At its center, the key rule directing energy inverters lies in their capacity to adjust the progression of power. This change, achieved through electronic circuits, empowers the incorporation of different energy sources into the power framework. This thorough investigation plans to dig into the complexities of energy inverters, unwinding their sorts, working standards, applications, and the powerful scene of inverter innovation.

Energy inverters manifest in assorted kinds, each fastidiously intended to address explicit requirements and applications. Extensively characterized, these inverters fall into three fundamental classes: sine wave inverters, changed sine wave inverters, and square wave inverters.

Sine wave inverters, the primary classification, produce a result waveform that intently mirrors the unadulterated sinusoidal waveform of utility-provided AC power. This kind of inverter is especially preferred for applications requesting a great AC power supply. Gadgets like touchy hardware, clinical gear, and certain modern apparatus benefit from the smooth and consistent waveform created by sine wave inverters. This trademark guarantees similarity, limiting the gamble of electrical obstruction or harm to associated gadgets.

Changed sine wave inverters, the subsequent classification, produce an air conditioner yield that approximates a ventured waveform. Albeit more affordable than sine wave inverters, they may not be appropriate for gadgets with explicit power necessities. Changed sine wave inverters find normal utilization in applications where cost is a basic element and the gadgets being controlled can endure a less refined AC waveform.

Square wave inverters comprise the third classification, creating a result waveform that looks like a square waveform. This kind of inverter is the most affordable but at the same time is the most un-normal because of its restricted applications. Square wave inverters are regularly utilized for fueling fundamental machines and instruments, yet their use is by and large stayed away from for gadgets that require a more steady and refined power source.

Understanding the functioning standards of energy inverters is critical for getting a handle on their usefulness and proficiency in switching DC over completely to AC power. The activity of inverters includes a few key stages, with the essential spotlight on DC-to-AC transformation. This change is accomplished using electronic circuits, frequently made out of semiconductors or other semiconductor gadgets, to switch the DC input in a way that duplicates the ideal AC waveform.

Beat Width Balance (PWM) stands apart as a usually utilized procedure in present day inverters. It includes fluctuating the width of the beats in the altered waveform to control the typical power conveyed to the heap. By changing the beat width, inverters can control the voltage and recurrence of the air conditioner yield, guaranteeing similarity with various kinds of burdens. This powerful control component takes into consideration productive energy change and flexibility to shifting working circumstances.

Energy inverters additionally consolidate control and criticism systems to keep up with steadiness and exactness in their result. These frameworks continually screen the result waveform, making ongoing changes in accordance with make up for varieties in burden or info conditions. This steady carefulness guarantees a predictable and solid AC power supply to associated gadgets, adding to the general dependability of the power framework.

In certain inverters, particularly those intended for high-power applications, transformer stages might be incorporated to additionally refine the result voltage.

Transformers help with changing the voltage levels and segregating the result from the information, giving extra assurance and upgrading the general exhibition of the inverter. This reconciliation of transformer stages is especially critical in applications where accuracy and unwavering quality are central.

Energy inverters track down applications across a different range of enterprises and settings, assuming a critical part in the productive use of electrical power. Remarkable applications remember their utilization for sustainable power frameworks, for example, sun oriented and wind power. In sunlight based establishments, inverters are instrumental in changing over the DC power produced by sun powered chargers into AC power for mix into the lattice or for direct use in homes and organizations. Essentially, wind turbines, which produce variable DC yield, depend on inverters to supply guarantee a steady AC.

In private settings, inverters are regularly utilized related to sunlight based chargers to change over DC power for family utilization. This permits mortgage holders to bridle sunlight based energy and lessen their reliance on customary matrix power, adding to supportability endeavors. Uninterruptible Power Supplies (UPS) likewise depend on inverters to give a consistent change from lattice capacity to battery power during blackouts. Without network power, the inverter changes over the DC power put away in batteries into AC power, guaranteeing a consistent and continuous power supply to basic gadgets and frameworks.

In modern applications, inverters assume an essential part in the activity of engine drives, control frameworks, and hardware. They consider exact control of engine speed and force, adding to energy productivity and cycle advancement in different modern cycles. The mix of energy inverters in electric vehicles (EVs) is another vital application. In this specific situation, inverters convert the DC power put away in the vehicle's batteries into AC ability to drive the electric engine, affecting the general proficiency and execution of electric vehicles.

Media communications address one more field where inverters are essential parts, particularly in reinforcement power frameworks. During blackouts, inverters guarantee a consistent progress to battery power, empowering the constant activity of basic correspondence hardware. Past these applications, energy inverters are fundamental in crisis and remote power situations, where admittance to framework power might be restricted. Versatile inverters can change over DC power from batteries, generators, or sunlight based chargers into AC power for assorted applications, including setting up camp, crisis lighting, and distant building destinations.

The field of inverter innovation has gone through critical progressions, driven by the rising interest for proficient and dependable power transformation. Eminent patterns and advancements in inverter innovation incorporate continuous endeavors to further develop proficiency.

Higher productivity not just diminishes energy misfortunes during the change cycle yet additionally adds to the general maintainability of influence frameworks. This drive toward upgraded effectiveness lines up with the more extensive objectives of energy preservation and ecological obligation.

Scaling down and joining of inverter parts address one more remarkable pattern in mechanical progression. Propels in semiconductor innovation have prompted more reduced and lightweight inverters, growing their reasonableness for a more extensive scope of uses. This pattern is especially obvious in compact gadgets and electric vehicles, where size and weight contemplations are basic.

The incorporation of shrewd advances, like microcontrollers and computerized signal processors, has led to savvy inverters. These high level inverters can speak with different gadgets, adjust to changing burden conditions, and give continuous information on energy creation and utilization. The joining of savvy highlights upgrades the general usefulness and versatility of inverters in assorted applications.

Matrix intuitive inverters have acquired unmistakable quality with the rising joining of environmentally friendly power sources into the framework. These inverters can synchronize with the network, taking care of abundance energy back into the framework and supporting matrix security. The capacity of lattice intelligent inverters to consistently coordinate environmentally friendly power into existing power foundation is urgent for progressing toward a more supportable and various energy blend.

High level control calculations assume an essential part in current inverters, empowering enhanced execution and versatility to dynamic working circumstances. These calculations work with fast reaction to changes in load, guaranteeing a steady and solid power supply. The persistent refinement of control calculations adds to the general proficiency and adequacy of energy inverters across different applications.

The continuous investigation of materials science and designing has prompted the advancement of inverters with further developed dependability and strength. These progressions are especially huge in applications where inverters are exposed to cruel natural circumstances or experience continuous cycling. The solidness of inverters is pivotal for keeping up with long haul execution and limiting the requirement for incessant support.

Staggered inverters have acquired notoriety in high-power applications. These inverters use various voltage levels to produce a flight of stairs like waveform, diminishing symphonious mutilation and working on the nature of the result waveform. The execution of staggered inverters is particularly helpful in applications where keeping a top notch AC power supply is fundamental.

The coordination of inverters with energy capacity frameworks, like batteries, has turned into a pervasive pattern. This combination takes into consideration better administration of energy stream, empowering the capacity of abundance energy during times of low interest and its delivery during top interest. The collaboration among inverters and energy stockpiling adds to by and large lattice soundness and unwavering quality.

While energy inverters have taken amazing steps as far as productivity and usefulness, a few difficulties and potential open doors lie ahead. As the portion of environmentally friendly power sources in the power matrix keeps on developing, challenges connected with framework coordination might emerge. Shrewd inverters outfitted with network steady elements are urgent for keeping up with matrix security and unwavering quality.

Coordinating inverters with energy capacity frameworks presents both specialized and monetary difficulties. Creating practical and productive answers for putting away and recovering energy stays a center region for future examination. Normalization of inverter innovations and guaranteeing interoperability among various frameworks are fundamental for making a consistent and interconnected energy scene. Vast norms can work with the joining of assorted energy sources and capacity frameworks.

With the rising dependence on computerized innovations, network protection is a developing worry in the energy area. Inverters, being basic parts in power frameworks, need vigorous network safety measures to safeguard against expected dangers and weaknesses. Proceeded with headways in semiconductor materials, for example, wide-bandgap semiconductors, hold the possibility to additionally work on the productivity and execution of energy inverters. These materials offer benefits like higher temperature resistance and diminished energy misfortunes.

The combination of numerous energy sources and capacity advancements in half and half energy frameworks presents energizing open doors for inverter innovation. Crossover frameworks that join sunlight based, wind, and other sustainable sources with conventional power sources require flexible and versatile inverters. With regards to worldwide jolt drives, energy inverters assume a critical part in stretching out admittance to power in remote and underserved locales. Creating practical and solid inverters custom-made to the particular requirements of these locales is really important.

As the quantity of inverters being used increments, tending to end-of-life issues and reusing contemplations becomes principal. Reasonable practices in assembling, reusing, and removal of inverters add to limiting natural effect. All in all, energy inverters stand as essential parts in current power frameworks, empowering the consistent reconciliation of different energy sources and working with the proficient use of electrical power.

From private sun powered establishments to modern applications and electric vehicles, the effect of inverters on our regular routines is significant. The advancement of inverter innovation keeps on being formed by the quest for higher effectiveness, brilliant usefulness, and supportability. As we explore the intricacies of an inexorably energized world, the job of energy inverters will without a doubt stay fundamental to the eventual fate of force age and utilization.

1.2 Brief historical overview of energy inverters.

The verifiable direction of energy inverters is set apart by an intriguing development, mirroring the unique exchange of innovative headways, modern requirements, and the consistently developing interest for proficient power change. The foundations of energy reversal can be followed back to the mid twentieth hundred years, agreeing with the quick improvement of substituting current (AC) power frameworks.

The spearheading work of creators and architects laid the basis for the rise of energy inverters. In the late nineteenth and mid twentieth hundreds of years, as AC power frameworks acquired noticeable quality over direct current (DC) frameworks, the need emerged for gadgets equipped for switching DC over completely to AC. This shift was driven by the innate benefits of AC, for example, the capacity to send control over longer distances with decreased energy misfortunes.

One of the earliest and most powerful figures in the improvement of energy inverters was Charles Proteus Steinmetz, a German-conceived American specialist and mathematician. Steinmetz, working for General Electric (GE) in the mid twentieth hundred years, made critical commitments to the comprehension of AC power and the improvement of AC-related advancements. His work established the hypothetical starting point for the plan and activity of early energy inverters.

The primary functional utilization of energy reversal arose with regards to early radio correspondence frameworks. In the mid 1920s, radio transmitters principally worked on DC power, yet the requirement for significant distance transmission prompted the improvement of inverters to switch DC over completely to AC. These early inverters were in many cases straightforward, using electromechanical parts to create an air conditioner yield.

As innovation progressed, so did the refinement of energy inverters. The mid-twentieth century saw the rise of strong state gadgets, making ready for the improvement of additional productive and minimized inverters. The approach of semiconductors and other semiconductor gadgets empowered the plan of inverters fit for switching DC over completely to AC with more noteworthy accuracy and dependability.

During the 1950s and 1960s, with the ascent of room investigation, inverters became vital parts of shuttle power frameworks. The capacity to change over DC power from batteries or sunlight

based chargers into AC power was significant for fueling different electronic frameworks on shuttle. This period saw the advancement of inverters with further developed effectiveness and dependability to meet the requesting prerequisites of room missions.

The 1970s denoted a huge defining moment for energy inverters with the developing interest in environmentally friendly power sources, especially sun based power. As sunlight based photovoltaic (PV) innovation built up some momentum, the requirement for inverters fit for changing over the DC power created by sunlight powered chargers into usable AC power became obvious. Early sun based inverters were simple and frequently had restricted effectiveness, however they established the groundwork for the sun oriented power frameworks that would become boundless in the next many years.

The late twentieth century saw a flood in the reception of energy inverters for different applications, driven by progressions in power gadgets and a developing accentuation on energy proficiency. Inverters became vital parts in uninterruptible power supply (UPS) frameworks, guaranteeing a consistent progress from network capacity to battery power during blackouts. This period additionally saw the development of high-recurrence exchanging inverters, which added to enhancements in proficiency and diminished size.

The turn of the 21st century achieved a reestablished center around environmentally friendly power and the combination of conveyed energy assets into the power matrix. This change in accentuation moved the improvement of lattice tied inverters intended to synchronize with the framework and feed overabundance energy back into the power dispersion organization. These network intelligent inverters assumed a significant part in the development of sunlight based power establishments and different types of circulated age.

The advancement of energy inverters went on with headways in semiconductor innovation, considering the improvement of additional refined and adaptable inverters. Computerized signal processors (DSPs) and microcontrollers became indispensable to inverter configuration, empowering exact control calculations and improved versatility to shifting working circumstances. Shrewd inverters outfitted with correspondence capacities and framework steady elements arose, giving a more unique and intuitive job in power frameworks.

As of late, the push towards manageability and the rising infiltration of sustainable power sources have sped up the improvement of energy stockpiling frameworks. Energy inverters have become key parts in these frameworks, working with the consistent combination of energy stockpiling gadgets like batteries. Inverter advancements that empower bidirectional power stream — charging the batteries during times of low interest and releasing during top interest — have become instrumental in enhancing energy use and network security.

The approach of electric vehicles (EVs) has additionally moved the development of energy inverters. In EVs, inverters assume a focal part in changing over DC power from the vehicle's batteries into AC ability to drive the electric engine. The proficiency and execution of these inverters straightforwardly influence the general productivity and scope of electric vehicles, making them basic parts in the change to electric transportation.

Looking forward, the authentic direction of energy inverters guides towards proceeded with advancement and variation toward the changing scene of force age and utilization. As the world looks to progress to cleaner and more maintainable energy sources, energy inverters will assume an undeniably significant part in coordinating sustainable power into the network, supporting energy stockpiling arrangements, and driving the up and coming age of electric vehicles.

Difficulties and open doors lie not too far off, including the requirement for normalization, network safety measures, and the improvement of inverters customized to explicit applications and locales. The authentic excursion of energy inverters reflects innovative headways as well as the more extensive cultural movements towards cleaner and more proficient energy frameworks. Before very long, energy inverters are ready to stay at the front of molding the fate of force frameworks, adding to a more economical and interconnected energy scene.

1.3 The role of energy inverters in transforming and optimizing energy.

Energy inverters, as extraordinary gadgets in the domain of electrical designing, assume a focal part in changing over and enhancing energy for a bunch of utilizations. The groundbreaking idea of energy reversal is clear in its capacity to change over direct ebb and flow (DC) into rotating momentum (AC), empowering the consistent combination of assorted energy sources into current power frameworks. This thorough investigation dives into the complex job of energy inverters, clarifying their importance in the change and advancement of energy across different areas.

At its center, the job of energy inverters is attached in their ability to overcome any barrier between various types of electrical power — explicitly, the change of DC to AC. This transformation is urgent in light of multiple factors, remarkably for working with the reconciliation of environmentally friendly power sources, guaranteeing similarity with existing power lattices, and enhancing the use of electrical power in assorted applications.

Environmentally friendly power sources, for example, sun based and wind power, dominatingly produce DC power. Sunlight powered chargers produce DC power straightforwardly from daylight through the photovoltaic impact, while wind turbines create variable DC yield from the rotational movement of their sharp edges. In the two cases, the innate idea of these sustainable sources requires the utilization of energy inverters to change over the created DC power into AC power reasonable for dissemination and utilization.

The joining of environmentally friendly power into power lattices is a groundbreaking part of energy reversal. Lattice tied inverters, intended to synchronize with the framework, empower the consistent infusion of environmentally friendly power into existing power foundation. This network communication is fundamental for adjusting energy organic market, improving framework solidness, and supporting the progress towards a more reasonable energy blend. As sustainable power establishments multiply, the job of energy inverters as empowering influences of this progress turns out to be progressively articulated.

Besides, energy inverters add to the advancement of energy use by guaranteeing the quality and similarity of the air conditioner power they produce. Sine wave inverters, specifically, produce a smooth and consistent AC waveform that intently mirrors the unadulterated sinusoidal waveform of utility-provided power. This top notch yield is significant for applications requiring accuracy and unwavering quality, like touchy electronic gadgets, clinical gear, and certain modern hardware.

In modern settings, the job of energy inverters reaches out to the improvement of engine drives and control frameworks. Inverters work with exact control of engine speed and force, adding to energy productivity and cycle streamlining. Variable Recurrence Drives (VFDs), a kind of inverter, consider dynamic acclimations to engine speed, bringing about energy reserve funds and further developed execution in different modern cycles.

Uninterruptible Power Supplies (UPS) further represent the streamlining part of energy inverters. In case of a blackout, UPS frameworks depend on inverters to consistently progress from matrix capacity to battery power, guaranteeing continuous activity of basic gadgets and frameworks. This ability is vital in situations where coherence of force is fundamental, for example, in server farms, medical care offices, and media communications foundation.

The coming of brilliant framework advancements further enhances the job of energy inverters in improving energy dissemination and utilization. Shrewd inverters, outfitted with cutting edge control calculations and correspondence abilities, can effectively partake in lattice the executives. They can answer continuous matrix conditions, change yield boundaries, and add to network soundness by offering subordinate types of assistance, for example, recurrence guideline and responsive power support.

With regards to private power frameworks, energy inverters matched with energy capacity arrangements add to the streamlining of energy use. Battery inverters empower the charging and releasing of batteries, permitting clients to store overabundance energy during times of low interest and delivery it during top interest. This unique energy the executives upgrades self-utilization of sun based produced power, lessens dependence on the framework, and gives a level of energy freedom to mortgage holders.

Electric vehicles (EVs) address another boondocks where energy inverters assume a crucial part in energy change and improvement. In EVs, inverters convert DC power from the vehicle's batteries into AC ability to drive the electric engine. The productivity and execution of these inverters straightforwardly influence the general effectiveness and scope of electric vehicles. Regenerative stopping mechanisms, worked with by inverters, further advance energy use by changing over dynamic energy back into electrical energy during slowing down, in this way expanding by and large vehicle effectiveness.

As innovation propels, the job of energy inverters in energy enhancement keeps on developing. The coordination of energy stockpiling frameworks, like lithium-particle batteries, with inverters has turned into a key center region. In this arrangement, inverters work with bidirectional power stream, considering the proficient charging and releasing of batteries. This mix is key to the idea of dispersed energy assets and microgrids, where nearby energy age, stockpiling, and utilization are facilitated for most extreme effectiveness and versatility.

The job of energy inverters in energy improvement isn't bound to fixed applications. Convenient inverters, normally utilized in outside exercises, setting up camp, and crisis circumstances, give a flexible answer for in a hurry power needs. These inverters can change over DC power from batteries, generators, or sunlight based chargers into AC power, enabling clients to charge electronic gadgets, run apparatuses, and enlighten spaces in remote or off-framework conditions.

Besides, progressions in semiconductor innovation have prompted the scaling down and combination of inverter parts. Versatile and conservative inverters are currently more open, permitting clients to tackle the advantages of energy reversal in various settings. This democratization of inverter innovation adds to the streamlining of energy access and utilization on a worldwide scale.

The continuous advancement of savvy urban communities and wise foundation additionally highlights the groundbreaking job of energy inverters. Inverters coordinated into building energy the board frameworks can upgrade the utilization of sun oriented power, battery capacity, and network power in view of ongoing interest and valuing conditions. This degree of mechanization and advancement holds the possibility to altogether diminish energy utilization, improve lattice strength, and add to supportability objectives.

Notwithstanding the groundbreaking and improving abilities of energy inverters, difficulties and contemplations endure. The changeability of sustainable power sources presents difficulties with regards to matrix incorporation and solidness. As the portion of sun based and wind power increments, inverters should adjust to fluctuating circumstances and add to network soundness through cutting edge control instruments.

Normalization and interoperability are pivotal perspectives that warrant consideration. Laying out extensive principles for energy inverters guarantees similarity among various frameworks and advances a strong and interconnected energy scene. This normalization is especially significant as the energy area goes through quick changes with the mix of different energy sources and capacity innovations.

Network safety arises as a basic concern, particularly as inverters become more interconnected and dependent on computerized innovations. Vigorous network protection measures are basic to shield energy foundation from potential digital dangers, guaranteeing the trustworthiness and unwavering quality of energy frameworks.

As the worldwide local area endeavors to address environmental change and progress towards a feasible energy future, the job of energy inverters will proceed to develop and grow. The enhancement of energy utilization, incorporation of inexhaustible sources, and headway of brilliant innovations will stay at the front of this advancement. The continuous innovative work in materials science, power hardware, and matrix the executives will additionally improve the capacities of energy inverters, empowering them to assume an essential part in molding the fate of energy change and streamlining.

Chapter 2

Basics of Energy Conversion

Energy change is an essential idea that assumes a vital part in different parts of our day to day routines and the working of the regular world. It alludes to the most common way of changing one type of energy into another, an omnipresent peculiarity that happens at various scales and in different settings. Understanding the fundamentals of energy change is significant for getting a handle on the standards basic the activity of incalculable gadgets, frameworks, and normal cycles.

At its center, energy change is administered by the laws of thermodynamics, which lay out the major guidelines that oversee the change of energy. These regulations give a structure to understanding the impediments and conceivable outcomes of energy transformation processes. The primary law of thermodynamics, otherwise called the law of preservation of energy, declares that energy can't be made or annihilated; it can change structures. This primary standard highlights the interconnected idea of energy and its capacity to change between different states.

The second law of thermodynamics presents the idea of entropy, a proportion of the issue or haphazardness in a framework. As per this regulation, in any energy transformation process, the all out entropy of a secluded framework will in general increment over the long haul. This infers that while energy is moderated, its quality will in general debase as it changes from additional coordinated, concentrated structures to scattered, less usable structures. The subsequent regulation likewise leads to the idea of irreversibility, proposing that numerous energy transformation processes are intrinsically irreversible, prompting a net expansion in entropy.

One of the most natural types of energy change is the change of mechanical energy into electrical energy as well as the other way around. This cycle is exemplified by generators and electric engines, which assume urgent parts in different enterprises and our regular day to day existences. Generators convert mechanical energy, frequently got from the movement of turbines or motors, into electrical energy. This change depends on the standard of electromagnetic acceptance, found by Michael Faraday in the nineteenth hundred years. Faraday's regulation expresses that a changing attractive field prompts an electromotive power (EMF) in a close by transmitter, prompting the age of electric flow.

Alternately, electric engines use the collaboration between attractive fields and electric flows to deliver mechanical movement. At the point when a current-conveying guide is set in an attractive field, it encounters a power, bringing about rotational movement. This rule, known as the Lorentz force, shapes the reason for the activity of electric engines, which are indispensable to various applications, from home devices to modern hardware.

Nuclear power, or intensity, addresses one more typical type of energy subject to change. The field of thermodynamics dives into the standards administering the change of intensity into different types of energy as well as the other way around. Heat motors, for example, those found in cars and power plants, represent the change of nuclear power into mechanical work. These motors work based on the Carnot cycle, a hypothetical system that lays out the most extreme effectiveness feasible in an intensity motor working between two temperature supplies.

During the Carnot cycle, heat is ingested from a high-temperature supply, causing the functioning substance (usually a gas) to grow and perform work. The substance then delivers intensity to a low-temperature supply, going through a pressure interaction. This cyclic arrangement of occasions addresses the key interaction behind the activity of intensity motors. While real motors veer off from the glorified Carnot cycle, understanding this model gives experiences into the elements affecting the proficiency of energy transformation in useful frameworks.

In the domain of sustainable power, photovoltaic cells offer a particular type of energy change by straightforwardly changing daylight into power. This cycle, known as the photovoltaic impact, depends on semiconductors, commonly made out of materials like silicon.

At the point when photons from daylight strike the semiconductor material, they energize electrons, creating an electric flow. Photovoltaic cells have become basic to sun oriented power frameworks, adding to the developing movement towards practical energy sources.

Energy change isn't restricted to designed frameworks; it is a key part of natural cycles too. Living organic entities continually participate in the transformation of energy to do fundamental capabilities. Cell breath, for example, includes the transformation of compound energy put away in supplements into adenosine triphosphate (ATP), a type of energy that cells can promptly utilize. This unpredictable cycle happens in mitochondria and fills in as a foundation of cell energy digestion.

Furthermore, photosynthesis addresses a surprising illustration of energy change in the plant realm. During photosynthesis, plants outfit sunlight based energy to change over carbon dioxide and water into glucose, a cycle that supports their development and gives energy to different physiological exercises. The perplexing biochemical pathways associated with photosynthesis highlight the tastefulness and proficiency of normal energy transformation processes.

The idea of energy change stretches out past the domains of material science and science to the universe of data innovation. In the domain of registering, parallel digits, or pieces, address a type of data that can be controlled and handled. The energy utilization related with data handling has turned into a basic thought in the plan and improvement of PC frameworks.

One of the foundations of data hypothesis is the idea of reversible calculation, where calculation can, in principle, be performed with zero energy dissemination. While useful executions of reversible calculation face critical difficulties, the quest for energy-productive figuring has prompted advancements, for example, quantum registering, which investigates the standards of quantum mechanics for data handling.

Understanding the nuts and bolts of energy change is fundamental for tending to the difficulties and potential open doors related with the worldwide energy scene. As social orders endeavor to change towards more feasible and harmless to the ecosystem energy sources, an exhaustive handle of energy transformation standards becomes central. Sustainable power advancements, like breeze turbines and sunlight based chargers, depend on successful energy transformation cycles to outfit regular assets and limit ecological effect.

Wind energy transformation includes catching the dynamic energy of moving air masses and changing it into electrical power. Wind turbines, outfitted with cutting edges that pivot because of wind, drive generators to create power. The proficiency of wind energy change relies upon elements, for example, wind speed, turbine plan, and the geographic area of wind ranches. Saddling wind power adds to enhancing the energy blend and lessening reliance on petroleum derivatives.

Likewise, hydropower addresses a long-laid out type of energy transformation, where the gravitational possible energy of water is changed over into power. Hydroelectric power plants influence the progression of water to turn turbines, producing electrical power. The versatility of hydropower, going from limited scope establishments to enormous dams, makes it a flexible and solid wellspring of sustainable power.

The developing interest in energy capacity advancements is firmly connected to the difficulties presented by the discontinuous idea of some sustainable power sources. Energy change assumes a significant part away frameworks that plan to store overabundance energy for sometime in the future. Battery innovations, going from customary lead-corrosive batteries to cutting edge lithium-particle batteries, work with the change among electrical and compound energy for productive capacity and recovery.

Past regular batteries, arising advancements, for example, siphoned hydro capacity and nuclear power stockpiling offer creative ways to deal with store and delivery energy. Siphoned hydro capacity includes siphoning water to a raised supply during times of abundance energy and delivering it to produce power during top interest. Nuclear power stockpiling uses stage change materials or intensity stockpiling frameworks to store and delivery nuclear power depending on the situation.

The field of energy change likewise converges with materials science, as the effectiveness and execution of energy transformation gadgets rely upon the properties of the materials in question. Propels in materials plan and designing add to the improvement of additional productive sun based cells, lightweight and strong breeze turbine parts, and superior execution anodes for batteries.

Nanotechnology, specifically, holds guarantee for changing energy transformation advancements by empowering exact command over materials at the nanoscale. Nanomaterials display novel properties that can upgrade the proficiency of sun oriented cells, work on the synergist movement of power modules, and enhance the presentation of thermoelectric gadgets. The coordination of nanotechnology into energy change processes addresses an outskirts in innovative work, offering possible leap forwards in effectiveness and supportability.

As the world wrestles with the difficulties of environmental change and endeavors to diminish ozone depleting substance emanations, the significance of energy transformation with regards to clean energy advances turns out to be progressively obvious. The progress from petroleum product subordinate frameworks to sustainable and reasonable energy sources requires headways in energy change proficiency, energy capacity, and the improvement of creative arrangements.

The job of energy transformation in transportation is a basic part of the more extensive energy scene. The customary gas powered motor, pervasive in vehicles for north of a really long period, changes over compound energy put away in petroleum products into mechanical work. The natural effect of burning based transportation has provoked a shift towards electric vehicles (EVs), where the energy transformation process includes the utilization of electric engines fueled by batteries.

EVs address a change in perspective in the car business, with an emphasis on diminishing discharges and dependence on limited petroleum derivative assets. The energy change effectiveness of electric vehicles, incorporating the age, transmission, and usage of electrical power, has suggestions for generally ecological supportability. The combination of sustainable power sources into the power matrix further upgrades the ecological advantages of electric transportation.

Energy transformation likewise assumes a significant part in the investigation and usage of space. Space missions, whether monitored or automated, depend on complex frameworks for power age, life backing, and drive. Sunlight based chargers are regularly utilized in space investigation to change over daylight into power, giving a solid power source to rocket and satellites. High level impetus frameworks, for example, particle drives, use electrical energy to create push for interplanetary travel.

The quest for proficient and supportable energy transformation innovations is firmly connected to the more extensive field of energy strategy and financial matters. Policymakers, industry pioneers, and scientists team up to characterize systems that advance the reception of clean energy innovations, address energy security concerns, and cultivate development. Financial variables, including the expense of energy creation, stockpiling, and circulation, impact the plausibility and versatility of various energy transformation arrangements.

The advancement of energy transformation innovations is interwoven with cultural movements, social perspectives, and international contemplations. As the worldwide local area wrestles with the basic to relieve environmental change and progress towards a more feasible energy future, the job of energy transformation in forming this direction couldn't possibly be more significant. Developments in energy change can possibly reshape enterprises, set out new monetary open doors, and add to a stronger and earth cognizant worldwide energy framework.

2.1 Explain the fundamental principles of energy conversion.

Energy change is a complicated and diverse interaction that lies at the core of innumerable normal peculiarities, innovative progressions, and ordinary exercises. At its center, energy change alludes to the change of one type of energy into another, a general rule that oversees the way of behaving of the actual world. To grasp the basic standards of energy change, one should dive into the domain of thermodynamics, the part of physical science that arrangements with the standards administering energy and its changes.

The primary law of thermodynamics, frequently alluded to as the law of preservation of energy, is a central rule that underlies all energy change processes. This regulation expresses that energy can't be made or obliterated; rather, it can change structures. The all out energy of a disengaged framework stays consistent after some time. This suggests that in any energy transformation process, the underlying energy should be represented in the last state, underlining the interconnected idea of various types of energy.

As energy goes through change, it advances between different states, each described by unambiguous properties. These states incorporate active energy, the energy related with movement, and likely energy, the energy put away in an article in light of its situation or setup. Mechanical energy, a mix of dynamic and possible energy, is much of the time a vital participant in energy transformation processes, especially in machines and mechanical frameworks.

The second law of thermodynamics presents the idea of entropy, a proportion of the problem or irregularity in a framework. This regulation directs that the all out entropy of a disengaged framework will in general increment over the long run. In basic terms, it proposes that cycles in the normal world will generally move towards conditions of more noteworthy issue. While the main regulation guarantees the preservation of energy, the subsequent regulation presents the idea that the nature of energy frequently corrupts during change, prompting a dispersal of energy and an expansion in entropy.

Irreversibility is a vital idea related with the second law of thermodynamics. Numerous energy change processes are intrinsically irreversible, implying that once energy goes through a change, it is testing or difficult to get back to its underlying state. This irreversibility adds to the general expansion in entropy, as energy scatters and turns out to be less amassed in additional usable structures.

Understanding the standards of energy change requires an investigation of different modes through which energy shows itself. Mechanical energy, as referenced prior, is predominant in the activity of machines and mechanical frameworks. The change among dynamic and potential energy is exemplified by a swinging pendulum, where the motor energy at the absolute bottom of the swing is step by step changed into expected energy at the most elevated point as well as the other way around.

Electromagnetic energy is one more huge type of energy that assumes a pivotal part in energy transformation processes. This incorporates noticeable light, radio waves, and different types of electromagnetic radiation. One striking illustration of electromagnetic energy change is the activity of sun based cells. Photovoltaic cells, usually found in sun powered chargers, convert daylight straightforwardly into electrical energy through the photovoltaic impact. This cycle includes the excitation of electrons in semiconductor materials by photons, bringing about the age of an electric flow.

Nuclear power, or intensity, is an unavoidable type of energy that emerges from the movement of iotas and particles. Heat move is a typical part of energy transformation, whether in the activity of motors, the age of power, or the warming of spaces. Nuclear power can be outfit to perform work, as exhibited in the activity of steam motors and gas turbines, where intensity is utilized to grow a functioning substance and drive mechanical movement.

Compound energy is put away in the connections among iotas and particles and can be delivered through substance responses. Ignition, the method involved with consuming fuel, is a natural illustration of compound energy change. In gas powered motors, for example, those in cars, substance energy put away in petroleum products is changed over into mechanical work and dynamic energy. The productivity of this change cycle is a critical thought in enhancing the presentation of motors and limiting waste.

Power is a flexible type of energy that considers effective transmission and usage in different applications. The standards of electromagnetic acceptance, found by Michael Faraday, structure the reason for the transformation among mechanical and electrical energy. Generators, for example, change mechanical energy, frequently got from the pivot of turbines or motors, into electrical energy by prompting an electromotive power (EMF) in a transmitter traveling through an attractive field.

Then again, electric engines work by changing over electrical energy into mechanical energy. The collaboration between attractive fields and electric flows brings about the age of mechanical movement. Electric engines are necessary to various gadgets and frameworks, from home devices to modern hardware, exhibiting the omnipresence and significance of electrical energy change in present day culture.

Environmentally friendly power innovations depend intensely on compelling energy change cycles to outfit regular assets and add to maintainable energy rehearses. Wind energy change includes catching the dynamic energy of moving air masses and changing it into power. Wind turbines, furnished with turning edges, create mechanical movement that is changed over into electrical power by generators. The effectiveness of wind energy transformation relies upon elements, for example, wind speed, turbine plan, and the geographic area of wind ranches.

Hydropower addresses one more longstanding type of energy change, where the gravitational expected energy of water is changed over into power. Hydroelectric power plants use the progression of water to turn turbines, producing electrical power. The adaptability of hydropower, from limited scope establishments to enormous dams, makes it a flexible and solid wellspring of environmentally friendly power. The standards administering the change of water's expected energy into electrical energy highlight the assorted manners by which energy transformation is utilized to address cultural issues.

Sun powered energy transformation, especially through photovoltaic cells, has acquired unmistakable quality in the mission for economical energy sources. The photovoltaic impact, which changes over daylight into power, depends on the semiconductor properties of materials like silicon. As photons from daylight strike the semiconductor, they energize electrons, producing an electric flow. Sunlight based chargers have become pervasive in applications going from private housetops to huge scope sun oriented ranches, adding to the worldwide shift towards cleaner and environmentally friendly power.

The effectiveness of energy transformation processes is a basic thought in the plan and streamlining of energy frameworks. The Carnot cycle, a hypothetical structure in thermodynamics, gives experiences into the greatest proficiency feasible in heat motors working between two temperature supplies. During the Carnot cycle, heat is ingested from a high-temperature repository, causing a functioning substance (ordinarily a gas) to extend and perform work. The substance then, at that point, discharges intensity to a low-temperature repository, going through a pressure interaction.

While genuine motors go astray from the romanticized Carnot cycle, understanding this model offers important benchmarks for surveying the exhibition of true frameworks. Reasonable contemplations, like grating, heat misfortunes, and irreversibilities, frequently lead to efficiencies lower than the Carnot effectiveness. Architects and analysts consistently endeavor to work on the productivity of energy transformation cycles to upgrade the supportability and financial practicality of energy advancements.

Energy capacity innovations assume a urgent part in tending to the irregularity of some environmentally friendly power sources and guaranteeing a solid and versatile energy supply. Batteries, specifically, work with the transformation among electrical and synthetic energy for productive capacity and recovery. Customary lead-corrosive batteries, as well as cutting edge lithium-particle batteries, are generally utilized for energy capacity in applications going from compact gadgets to electric vehicles and network scale energy capacity.

Siphoned hydro capacity addresses one more imaginative way to deal with energy capacity, including the development of water among raised and lower repositories to store and delivery energy. During times of abundance energy, water is siphoned to a raised supply, and during top interest, producing electricity is delivered. Nuclear power stockpiling uses materials with stage change properties or specific frameworks to store and delivery nuclear power on a case by case basis.

Materials science assumes a crucial part in propelling the proficiency and execution of energy change gadgets. The properties of materials impact the plan and activity of innovations like sun based cells, wind turbine parts, and batteries. Nanotechnology, specifically, holds guarantee for upsetting energy transformation by empowering exact command over materials at the nanoscale.

Nanomaterials show extraordinary properties that can upgrade the effectiveness of energy transformation processes. In sun oriented cells, for instance, nanomaterials can work on light assimilation and electron transport, prompting higher change efficiencies. Additionally, nanotechnology research adds to the improvement of cutting edge materials for lightweight and strong breeze turbine parts, as well as elite execution anodes for batteries and power modules.

The reconciliation of nanotechnology into energy change processes addresses a boondocks in innovative work, offering expected forward leaps in productivity and manageability. As specialists investigate novel materials and nanoscale designing methodologies, the journey for more productive and harmless to the ecosystem energy change advancements proceeds.

Natural frameworks give probably the most amazing instances of energy change. Living creatures continually participate in the change of energy to do fundamental capabilities. Cell breath, a central cycle in cells, includes the change of compound energy put away in supplements into adenosine triphosphate (ATP), a type of energy that cells can promptly utilize. This complicated cycle happens in mitochondria and fills in as a foundation of cell energy digestion.

Photosynthesis, happening in plants and certain microorganisms, is one more remarkable illustration of energy transformation. During photosynthesis, plants bridle sun oriented energy to change over carbon dioxide and water into glucose, a cycle that supports their development and gives energy to different physiological exercises. The complex biochemical pathways engaged with photosynthesis feature the tastefulness and productivity of regular energy change processes.

The field of energy transformation stretches out past the limits of material science and science to the domain of data innovation. In figuring, parallel digits, or pieces, address a type of data that can be controlled and handled. The energy utilization related with data handling has turned into a basic thought in the plan and enhancement of PC frameworks.

Data hypothesis presents the idea of reversible calculation, where calculation can, in principle, be performed with zero energy dispersal. While pragmatic executions of reversible calculation face critical difficulties, the quest for energy-effective processing has prompted developments, for example, quantum figuring. Quantum PCs influence the standards of quantum mechanics for data handling, offering the potential for fundamentally quicker and more energy-proficient calculation contrasted with traditional PCs.

The interdisciplinary idea of energy transformation research is apparent in its convergence with energy strategy and financial aspects.

Policymakers, industry pioneers, and specialists team up to characterize procedures that advance the reception of clean energy advances, address energy security concerns, and encourage development.

Monetary elements, including the expense of energy creation, stockpiling, and conveyance, impact the achievability and adaptability of various energy transformation arrangements.

The development of energy change advancements is firmly connected to cultural movements, social perspectives, and international contemplations. As the worldwide local area wrestles with the basic to moderate environmental change and progress towards a more maintainable energy future, the job of energy transformation in forming this direction turns out to be progressively obvious. Developments in energy change can possibly reshape enterprises, set out new monetary open doors, and add to a stronger and earth cognizant worldwide energy framework.

Transportation addresses a basic part of the more extensive energy scene, and the standards of energy change assume an essential part in this space. The customary gas powered motor, which has ruled auto innovation for north of 100 years, changes over synthetic energy put away in petroleum products into mechanical work. The natural effect of burning based transportation has provoked a shift towards electric vehicles (EVs), where the energy transformation process includes the utilization of electric engines controlled by batteries.

Electric vehicles address a change in perspective in the car business, with an emphasis on lessening outflows and dependence on limited non-renewable energy source assets. The energy transformation proficiency of electric vehicles, enveloping the age, transmission, and usage of electrical power, has suggestions for generally speaking ecological maintainability. The coordination of sustainable power sources into the power framework further improves the ecological advantages of electric transportation.

The standards of energy transformation stretch out to the investigation and usage of space. Space missions, whether monitored or automated, depend on refined frameworks for power age, life backing, and drive. Sun powered chargers are generally utilized in space investigation to change over daylight into power, giving a solid power source to space apparatus and satellites. High level impetus frameworks, for example, particle drives, use electrical energy to produce push for interplanetary travel.

As the world keeps on wrestling with the difficulties of environmental change, the significance of energy transformation with regards to clean energy advancements turns out to be progressively clear. The change from petroleum product subordinate frameworks to sustainable and reasonable energy sources requires progressions in energy transformation proficiency, energy capacity, and the advancement of creative arrangements. The standards of energy change act as a directing structure for tending to the worldwide energy emergency and cultivating an additional reasonable and versatile future.

2.2 Different types of energy sources and their conversion into electrical energy.

The variety of fuel sources accessible to humankind is huge, enveloping both sustainable and non-inexhaustible choices. These sources fluctuate in their starting point, attributes, and natural effect. Understanding the various kinds of fuel sources and how they can be changed over into electrical energy is urgent for molding a supportable and strong energy scene.

Petroleum derivatives, like coal, oil, and flammable gas, have generally been the essential wellsprings of electrical energy. The change interaction commonly includes consuming these energizes to deliver heat, which is then used to create steam. This steam, thusly, drives turbines associated with generators that produce power. While this strategy has been compelling, it accompanies critical ecological results, including air contamination, ozone depleting substance discharges, and reliance on limited assets.

Coal-terminated power plants, one of the most seasoned and most normal types of power age, depend on the ignition of coal to produce heat. The intensity delivered changes over water into steam, which drives turbines associated with generators. The rotational movement of the turbines produces electrical energy. Notwithstanding, the ecological effect of coal-terminated power plants, described by high carbon dioxide outflows and different poisons, has prompted expanded examination and a push towards cleaner choices.

Oil-put together power plants work with respect to a comparable guideline however use oil, for example, diesel or weighty fuel oil, as the essential fuel. These power plants are many times utilized in districts where admittance to coal is restricted. The burning of oil produces heat, which is utilized to make steam, at last driving turbines and creating power. In spite of their adaptability, oil-based power plants face difficulties connected with cost unpredictability, international worries, and natural effect.

Gaseous petrol power plants have acquired prominence due to their somewhat lower natural effect contrasted with coal and oil. Flammable gas, made chiefly out of methane, goes through burning to deliver heat, which is then used to produce steam. The following system of driving turbines and generators changes over the nuclear power into electrical energy. Petroleum gas power plants are known for their effectiveness and lower carbon dioxide discharges however are as yet connected with ecological difficulties, including methane spillage during extraction and transport.

Atomic power addresses one more huge wellspring of electrical energy, depending on the transformation of atomic splitting responses. In thermal energy stations, the core of a weighty particle, commonly uranium-235 or plutonium-239, goes through splitting, delivering a lot of intensity. This intensity is utilized to create steam, which drives turbines associated with generators.

Atomic power offers a high energy thickness and low ozone depleting substance emanations yet accompanies concerns connected with radioactive garbage removal, wellbeing, and the potential for atomic mishaps.

Environmentally friendly power sources have acquired conspicuousness as cleaner and more reasonable options in contrast to customary petroleum derivatives. Among these, sunlight based energy stands apart as a promising choice for electrical energy age. Photovoltaic cells, normally tracked down in sun powered chargers, straightforwardly convert daylight into power through the photovoltaic impact. At the point when photons from daylight strike the semiconductor material in the sun oriented cells, they energize electrons, making an electric flow. Sunlight based power frameworks bridle this electrical energy for different applications, from private roof establishments to enormous scope sun oriented ranches.

Wind energy is one more inexhaustible source that saddles the active energy of moving air masses to create power. Wind turbines, outfitted with turning cutting edges, catch the breeze's energy and convert it into rotational movement. The turning movement drives a generator, changing the mechanical energy into electrical energy. Wind power has turned into a huge supporter of worldwide power age, offering a spotless and bountiful asset, however difficulties, for example, irregularity and visual effect should be tended to.

Hydropower, one of the most established types of energy change, takes advantage of the gravitational expected energy of water to create power. In hydroelectric power plants, water is let out of a repository, moving through turbines that drive generators. The active energy of the moving water is changed over into mechanical energy and, hence, electrical energy. Hydropower is exceptionally dependable and versatile, with applications going from enormous dams to little run-of-stream establishments.

Geothermal energy takes advantage of the World's inward intensity as a wellspring of electrical power. This energy is outfit through geothermal power plants, where steam or high temp water from underground repositories is brought to the surface. The steam drives turbines, associated with generators, delivering power. Geothermal power is viewed as a spotless and persistent energy source, especially in districts with high geothermal action, like Iceland and portions of the US.

Biomass, got from natural materials like wood, crop buildups, and waste, addresses a flexible sustainable power source. The burning of biomass discharges heat, which can be utilized straightforwardly for warming or changed over into power through biomass power plants. In these power plants, biomass is singed to deliver steam, driving turbines associated with generators. While biomass is sustainable, its natural effect relies upon elements, for example, land use, development practices, and emanations from burning.

Sea energy envelops different advances that outfit the energy put away in seas and oceans. Flowing energy, created by the gravitational draw of the moon and the sun on Earth's seas, can be changed over into power utilizing flowing power plants. These plants utilize the rhythmic movement of tides to drive turbines and generators. Additionally, wave energy converters catch the dynamic energy of sea waves to produce power. Sea nuclear power transformation (OTEC) uses the temperature contrast between warm surface water and cold profound water to create steam and drive turbines.

The transformation of these different energy sources into electrical energy includes normal standards in light of electromagnetism and electromagnetic acceptance. In generators, the rotational movement of a loop or transmitter inside an attractive field prompts an electromotive power (EMF), making an electric flow. The rotating flow (AC) created in generators is then frequently switched over completely to coordinate flow (DC) for circulation and use in electrical matrices.

Transmission and conveyance networks assume a significant part in conveying electrical energy from power plants to end-clients. High-voltage transmission lines transport power over significant distances with insignificant misfortunes, while conveyance networks disperse influence to homes, organizations, and enterprises. Substations and transformers work with the transformation among high and low voltages, guaranteeing proficient and dependable power supply.

Energy capacity innovations address the discontinuity and inconstancy related with some environmentally friendly power sources, taking into consideration the capacity of abundance energy for sometime in the future. Batteries, for example, lithium-particle batteries, store electrical energy in synthetic structure and delivery it when required. Siphoned hydro capacity frameworks utilize surplus power to siphon water to a raised supply, delivering it later to create power during appeal periods. Nuclear power stockpiling includes putting away intensity or cold for sometime in the future, improving the adaptability of sustainable power mix.

As the worldwide local area wrestles with the difficulties of environmental change and looks to progress towards economical energy frameworks, the job of energy transformation becomes foremost. Propels in innovation, materials science, and energy strategy will keep on forming the scene of fuel sources and their change into electrical energy. The reconciliation of savvy matrices, energy-proficient advances, and imaginative stockpiling arrangements will add to a stronger and manageable energy future.

All in all, the variety of fuel sources accessible for electrical energy age traverses conventional petroleum derivatives, atomic power, and a developing set-up of sustainable choices. Each source has its interesting change processes and natural contemplations. Petroleum product put together power plants depend with respect to burning to deliver heat for steam turbines, while thermal energy stations bridle the energy delivered during atomic parting.

Interestingly, sustainable sources like sunlight based, wind, hydropower, geothermal, biomass, and sea energy utilize different advances to straightforwardly or in a roundabout way convert normal assets into electrical energy.

The standards of electromagnetic acceptance oversee the transformation of mechanical, warm, or synthetic energy into electrical energy in generators. The subsequent electrical energy is then communicated through high-voltage lines, disseminated through neighborhood networks, and put away utilizing progressed energy capacity advancements. As the world takes a stab at a more feasible and tough energy future, proceeded with innovative work in energy transformation advances, combined with vital strategy choices, will assume a significant part in molding the direction of worldwide energy frameworks.

2.3 Introduce the concept of inverter technology and its role in energy conversion.

Inverter innovation addresses a crucial headway in the field of energy change, assuming an extraordinary part in how electrical energy is produced, put away, and used. This innovation is especially conspicuous with regards to exchanging flow (AC) and direct flow (DC) frameworks, offering a flexible method for changing over between these two types of electrical energy. The presentation of inverter innovation has essentially affected different areas, from sustainable power frameworks and electronic gadgets to electric vehicles and modern applications.

At its center, an inverter is an electronic gadget intended to change over DC power into AC power or the other way around. This capacity to change the electrical qualities of a power source makes inverters vital in many applications where the similarity between various electrical frameworks is fundamental. Inverters act as an extension among DC and AC power, empowering consistent mix and effective usage of electrical energy in different settings.

In the domain of environmentally friendly power, inverters assume a significant part in the transformation of DC power created by sources, for example, sunlight based chargers and wind turbines into AC power reasonable for circulation in electrical lattices. Photovoltaic (PV) frameworks, which outfit daylight to produce DC power, use inverters to change this DC power into the air conditioner power utilized in homes and organizations. Additionally, wind turbines create variable-speed AC power, and inverters guarantee the synchronization and lattice similarity of the power they produce.

The productivity of sustainable power frameworks is enormously impacted by the exhibition of inverters. Present day network tied sunlight based inverters, for example, integrate progressed highlights, for example, greatest power point following (MPPT), which advances the result of sun powered chargers by changing the electrical working point. MPPT guarantees that the sunlight based chargers work at their most extreme effectiveness, upgrading the general energy yield of the framework.

Off-lattice and independent environmentally friendly power frameworks likewise vigorously depend on inverters for energy transformation. In these arrangements, inverters are fundamental for changing over DC power put away in batteries into AC power for home devices and other electrical gadgets. This is especially predominant in distant regions where admittance to a unified power matrix is restricted, and sun based or wind energy fills in as the essential power source.

Inverter innovation plays had an extraordinary impact in the jolt of transportation, especially in electric vehicles (EVs). EVs use high-voltage DC power put away in batteries for drive. Inverters in electric vehicles are answerable for changing over this DC power into AC ability to drive the electric engine. The capacity to proficiently change over power among DC and AC is basic for the exhibition, reach, and in general effectiveness of electric vehicles.

Moreover, regenerative slowing mechanisms in electric vehicles influence inverter innovation. During slowing down, the electric engine capabilities as a generator, changing over motor energy back into electrical energy. The inverter assumes a critical part in this cycle by changing the produced electrical energy from AC over completely to DC for capacity in the vehicle's battery. This regenerative slowing down ability improves the general energy effectiveness of electric vehicles.

The inescapable reception of inverter innovation altogether affects buyer gadgets and home devices. Numerous electronic gadgets, like PCs, cell phones, and home hardware, work on DC power. Inverters incorporated into power connectors and electronic gadgets convert AC power from outlets into the DC power expected for the gadgets to work. This guarantees a consistent point of interaction between the air conditioner power provided by electrical lattices and the DC power required by electronic contraptions.

With regards to cooling frameworks, the coming of inverter innovation has prompted critical headways in energy effectiveness. Customary forced air systems work on a fixed-speed blower, bringing about on-off cycles to keep up with the ideal temperature. Conversely, inverter forced air systems use variable-speed blowers driven by inverters. This takes into account exact control of the cooling limit, empowering the framework to change its result as per the cooling interest. Thus, inverter forced air systems are more energy-effective, give more steady solace, and add to energy investment funds.

Modern applications likewise benefit from the flexibility and productivity of inverter innovation. Variable recurrence drives (VFDs), a kind of inverter, are broadly utilized in modern engine control frameworks. VFDs control the speed and force of electric engines by changing the recurrence and voltage of the power provided to the engine. This capacity not just upgrades the energy productivity of modern cycles yet additionally takes into consideration better control and improvement of different tasks.

The idea of inverter innovation stretches out to uninterruptible power supply (UPS) frameworks, which are critical for giving reinforcement power in the event of electrical lattice disappointments. In UPS frameworks, inverters convert DC power put away in batteries into AC ability to guarantee a nonstop and stable power supply to basic hardware and gadgets. This application is especially fundamental in settings where continuous power is imperative, for example, server farms, medical clinics, and crisis administrations.

Microgrids, which are restricted energy frameworks fit for working freely or related to the really electrical matrix, frequently depend on inverter innovation for proficient energy change. In microgrids fueled by sustainable power sources and energy stockpiling, inverters work with the mix of different energy assets, guaranteeing a dependable and adjusted power supply to the neighborhood local area or office.

The development of inverter innovation has been set apart by headways in power gadgets, semiconductor gadgets, and control calculations. The effectiveness, unwavering quality, and in general execution of inverters have worked on fundamentally throughout the long term, adding to their boundless reception across different areas. Current inverters frequently consolidate highlights like high level control methodologies, constant checking, and correspondence capacities, empowering consistent coordination into brilliant frameworks and energy the executives frameworks.

Notwithstanding the various benefits of inverter innovation, certain difficulties endure. One critical test is the potential for electromagnetic impedance (EMI) and consonant mutilation brought by inverters into the electrical framework. These unsettling influences can influence the exhibition of other associated gadgets and compromise the nature of electrical power. Consequently, relieving EMI and symphonious bending is a basic thought in the plan and execution of inverter frameworks.

Moreover, the natural effect of the materials utilized in inverter producing and their finish of-life removal represents a test. Endeavors are in progress to foster more manageable and eco-accommodating inverter advances, consolidating recyclable materials and improving assembling cycles to limit natural mischief.

As the world keeps on changing towards a more economical and decentralized energy scene, the job of inverter innovation is set to turn out to be significantly more noticeable. The reconciliation of sustainable power sources, energy capacity frameworks, and electric vehicles into the standard energy foundation requires progressed and versatile energy transformation advances. Inverters, with their capacity to consistently change over between various electrical organizations, arise as a key part in this groundbreaking excursion towards a cleaner, more proficient, and decentralized energy future.

3

Chapter 3

Types of Energy Inverters

Energy inverters assume a vital part in present day power frameworks by changing over direct current (DC) into rotating current (AC) or the other way around. This change is fundamental for different applications, from controlling domestic devices to working with environmentally friendly power joining into the lattice. The sorts of energy inverters differ in view of their plan, usefulness, and expected use. In this far reaching investigation, we will dig into the universe of energy inverters, analyzing their arrangements, applications, and the advancing scene of energy transformation advancements.

One of the essential qualifications among energy inverters lies in their capacity to switch DC over completely to AC or AC to DC. Framework tie inverters, otherwise called network associated or matrix between tie inverters, are intended to change over DC power produced by sunlight based chargers or wind turbines into AC power that can be taken care of into the electrical lattice. These inverters synchronize with the network's recurrence and voltage, guaranteeing a consistent reconciliation of sustainable power sources into the current power foundation. The matrix attach inverter empowers clients to contribute overabundance energy back to the lattice, frequently through net metering game plans where the client is credited for the excess power provided.

Off-matrix inverters, then again, take care of independent power frameworks that are not associated with the utility network. These inverters convert DC power from batteries or sustainable sources, like sunlight based chargers or wind turbines, into AC power for use in far off areas or in crisis reinforcement frameworks. Off-matrix inverters are urgent for applications like remote lodges, boats, or crisis power supply during lattice blackouts. They are intended to deal with the variances in energy creation and utilization in segregated frameworks, guaranteeing a steady and solid power supply.

Inside the domain of matrix tie inverters, there are two primary classifications: string inverters and microinverters. String inverters are regularly utilized in private and business sun powered establishments. They are named for the strings of sunlight powered chargers they associate with, with the boards wired in series. In a string inverter framework, in the event that one board fails to meet expectations or is concealed, it influences the whole string's result. In any case, progressions in Greatest Power Point Following (MPPT) innovation have further developed string inverter productivity by improving the exhibition of each board separately.

Microinverters, then again, are little inverters introduced straightforwardly on each sun powered charger. This plan takes into account individual board improvement and limits the effect of concealing or breakdowns on the general framework execution. Microinverters are especially beneficial in establishments where boards might have fluctuating directions or are liable to concealing at various times. While string inverters are much of the time more practical in huge scope establishments, microinverters offer expanded adaptability and effectiveness in specific situations.

One more significant order of energy inverters depends on their waveform yield. The waveform alludes to the state of the air conditioner voltage delivered by the inverter. The two essential kinds of waveforms are unadulterated sine wave and adjusted sine wave.

Unadulterated sine wave inverters produce a smooth and steady waveform that intently impersonates the regular AC power given by the utility framework. Machines and electronic gadgets by and large work all the more effectively with unadulterated sine wave power. These inverters are reasonable for a great many applications, including delicate gadgets, engines, and machines. While unadulterated sine wave inverters are regularly more costly, their capacity to give spotless and stable power makes them fundamental in applications where accuracy and dependability are foremost.

Changed sine wave inverters, then again, create a ventured waveform that approximates a sine wave. While they are more practical than unadulterated sine wave inverters, they may not be appropriate for specific kinds of hardware. Gadgets with electric engines, for example, may encounter decreased proficiency and expanded heat age when controlled by a changed sine wave. Be that as it may, numerous domestic devices and instruments can work successfully with a changed sine wave power supply.

Notwithstanding waveform order, energy inverters can be sorted in light of their power limit. Inverter sizes are usually estimated in kilowatts (kW) or megawatts (MW) for bigger frameworks. Limited scope applications, for example, driving a PC or a cooler during a blackout, may require inverters in the scope of a couple hundred watts. Private sun based establishments ordinarily use inverters with limits going from 1 kW to 10 kW, contingent upon the size of the framework.

Business and modern applications frequently require bigger inverters, and utility-scale sun powered or wind ranches may send inverters with limits in the megawatt range. The power limit of an inverter is a critical consider deciding the general framework proficiency and the quantity of boards or sources it can oblige. Matching the inverter size to the particular necessities of the application is fundamental for ideal execution and cost-viability.

As the interest for sustainable power sources keeps on developing, the mix of energy stockpiling frameworks has become progressively predominant. Energy capacity permits abundance energy created during times of high creation to be put away for use during seasons of low creation or popularity. Inverter-chargers are gadgets that consolidate the elements of an inverter and a battery charger, permitting consistent reconciliation of energy stockpiling into a power framework.

Inverter-chargers are regularly utilized in off-matrix and half breed frameworks. In off-network establishments, the inverter-charger guarantees a steady AC power supply while likewise charging the batteries when an outer power source, like a generator or sunlight based chargers, is accessible. During times of low energy creation, the inverter draws power from the batteries to satisfy the heap need.

Mixture frameworks consolidate environmentally friendly power sources with network availability and energy stockpiling. In such frameworks, the inverter-charger deals with the progression of energy between the sustainable sources, the matrix, and the stockpiling framework. This adaptability empowers clients to improve their energy utilization in light of elements, for example, energy costs, matrix strength, and ecological contemplations.

Progressions in inverter innovation have prompted the improvement of savvy inverters, which consolidate correspondence and control abilities. Shrewd inverters empower continuous checking, controller, and matrix communication highlights. These inverters can speak with utility suppliers and network administrators to add to framework solidness, answer matrix signals, and improve the general proficiency and dependability of the power framework.

One vital element of brilliant inverters is their capacity to give network support capabilities, like voltage and recurrence guideline. This capacity becomes significant as the entrance of sustainable power sources increments, bringing inconstancy and discontinuity into the matrix. Shrewd inverters can assist with moderating these difficulties by effectively taking part in matrix the executives and offering subordinate types of assistance.

Notwithstanding matrix support, brilliant inverters add to the idea of a "savvy lattice." A shrewd network consolidates progressed correspondence and control innovations to upgrade the productivity, unwavering quality, and manageability of the electric power framework. Savvy inverters assume a crucial part in this progress by working with bidirectional correspondence between the utility and the disseminated energy assets, empowering a more powerful and responsive lattice foundation.

The idea of energy flexibility has acquired conspicuousness lately, determined by the need to guarantee a solid power supply despite catastrophic events, matrix disappointments, or different disturbances. Energy versatility includes the capacity of a framework to endure and recuperate from disturbances, guaranteeing progression of force supply for basic foundation and fundamental administrations. Inverters, particularly those coordinated with energy capacity frameworks, contribute altogether to improving energy flexibility.

Microgrid frameworks epitomize the use of energy versatility standards. A microgrid is a confined energy framework that can work freely or related to the principal network. It normally comprises of dispersed energy assets, like sunlight powered chargers, wind turbines, energy capacity, and high level inverters. In case of a framework blackout, a microgrid can keep on providing capacity to basic burdens, guaranteeing that fundamental administrations stay functional.

High level inverters inside microgrids empower consistent progress between network associated and islanded modes. Islanding alludes to the capacity of a microgrid to disengage from the principal network and work independently. In such situations, the inverters assume a pivotal part in keeping up with the harmony between energy age and utilization inside the microgrid, guaranteeing dependability and unwavering quality during detachment.

The ascent of electric vehicles (EVs) plays additionally extended the part of energy inverters. Electric vehicle charging stations frequently integrate inverters to change over lattice power into the DC power expected for charging EV batteries. Furthermore, bidirectional inverters empower vehicle-to-network (V2G) capacities, permitting EVs to release put away energy back to the lattice during top interest periods. This bidirectional progression of energy between the lattice and EVs adds to matrix soundness and gives a potential income stream to EV proprietors.

As innovation keeps on progressing, new materials and plans are being investigated to improve the proficiency and execution of energy inverters. Silicon carbide (SiC) and gallium nitride (GaN) are arising as elective semiconductor materials for power hardware. These materials offer higher warm conductivity and lower exchanging misfortunes contrasted with conventional silicon-based gadgets, bringing about additional productive and minimal inverters.

The journey for higher effectiveness and lower natural effect has additionally prompted the advancement of cutting edge inverters with further developed power thickness. High-recurrence exchanging, high level cooling methods, and imaginative circuit geographies add to decreasing the size and weight of inverters while keeping up with or working on their exhibition. Smaller and lightweight inverters are especially important in applications where space and weight limitations are basic, like electric vehicles and convenient power frameworks.

The coordination of man-made consciousness (computer based intelligence) and AI (ML) into energy the board frameworks is one more boondocks in inverter innovation. Man-made intelligence driven calculations can enhance the activity of inverters by anticipating energy creation designs, load requests, and matrix conditions. This prescient capacity empowers the inverter to proactively change its boundaries for ideal execution, adding to expanded productivity and lattice security.

Moreover, simulated intelligence and ML can improve shortcoming location and diagnostics in inverters. The capacity to recognize and resolve issues continuously further develops framework unwavering quality and diminishes margin time. This proactive support approach is especially significant in remote or unavailable places where manual examinations might challenge.

All in all, the universe of energy inverters is dynamic and complex, enveloping a different scope of advancements and applications. From matrix tie inverters working with the joining of sustainable power into the power framework to off-lattice inverters empowering energy freedom in far off areas, these gadgets assume a significant part in current energy frameworks. The development of inverter innovation, set apart by the approach of savvy inverters, energy capacity reconciliation, and high level materials, mirrors the continuous endeavors to improve proficiency, unwavering quality, and manageability in the domain of force transformation.

As we plan ahead, the proceeded with development of sustainable power, the zap of transportation, and the interest for strong power frameworks will drive further advancements in energy inverter plan and usefulness. The incorporation of computer based intelligence, ML, and high level materials will probably prepare for more clever, productive, and reduced inverters, adding to the continuous change of the worldwide energy scene. In this powerful and consistently evolving field, energy inverters will keep on being at the very front of empowering a practical and tough energy future.

3.1 Explore various types of energy inverters (sine wave, modified sine wave, grid-tie, off-grid, etc.).

Energy inverters are fundamental parts in power frameworks, working with the change of direct current (DC) to exchanging current (AC) or the other way around. The different utilizations of energy inverters range from fueling regular domestic devices to empowering the joining of sustainable power into the framework.

One major classification of energy inverters depends on their ability to change over DC power from various sources into AC power or the other way around. Network tie inverters, otherwise called lattice associated or framework between tie inverters, are intended to change over DC power produced by sources, for example, sunlight based chargers or wind turbines into AC power that can be taken care of into the electrical matrix.

The critical component of framework attach inverters is their capacity to synchronize with the lattice's recurrence and voltage, guaranteeing a consistent incorporation of sustainable power sources into the current power foundation. This synchronization permits clients to contribute abundance energy back to the framework, frequently prompting remuneration through net metering game plans where clients get credits for the excess power provided. Matrix tie inverters are vital for supporting the broad reception of sun based and wind energy by empowering a bi-directional progression of force between inexhaustible sources and the framework.

Conversely, off-network inverters are intended for independent power frameworks that work freely of the utility lattice. These inverters assume a basic part in changing over DC power from batteries or sustainable sources, like sun powered chargers or wind turbines, into AC power for use in distant areas or as crisis reinforcement frameworks. Off-framework frameworks are usually utilized in situations where lattice network is inaccessible or untrustworthy, like remote lodges, boats, or crisis power supply during matrix blackouts. Off-matrix inverters are designed to deal with the vacillations in energy creation and utilization inside separated frameworks, guaranteeing a steady and dependable power supply.

Inside the classification of lattice tie inverters, two fundamental subtypes are string inverters and microinverters. String inverters are generally utilized in private and business sun based establishments. They get their name from the strings of sunlight powered chargers they associate with, with these boards being wired in series. Notwithstanding, one limit of string inverters is that assuming one board fails to meet expectations or is concealed, it can influence the whole string's result. Propels in Most extreme Power Point Following (MPPT) innovation have tended to this constraint by enhancing the presentation of each board separately, working on the general proficiency of string inverter frameworks.

Microinverters, then again, address an alternate methodology. Rather than interfacing with a line of sunlight based chargers, microinverters are little inverters introduced straightforwardly on each sunlight powered charger. This plan considers individual enhancement of each board and limits the effect of concealing or breakdowns on the general framework execution. Microinverters are especially worthwhile in establishments where boards might have fluctuating directions or are liable to concealing at various times. While string inverters are in many cases more financially savvy in huge scope establishments, microinverters offer expanded adaptability and proficiency in specific situations.

One more critical arrangement of energy inverters depends on the waveform of the air conditioner yield they produce. Waveform alludes to the state of the air conditioner voltage, and the two essential sorts are unadulterated sine wave and altered sine wave inverters.

Unadulterated sine wave inverters produce a smooth and predictable waveform that intently looks like the regular AC power given by the utility network. This perfect and stable power yield is fundamental for the proficient activity of a great many gadgets, including delicate hardware, engines, and machines. While unadulterated sine wave inverters are commonly more costly than their adjusted sine wave partners, their capacity to give top notch power makes them crucial in applications where accuracy and dependability are principal.

Conversely, changed sine wave inverters produce a ventured waveform that approximates a sine wave. These inverters are more financially savvy however may not be appropriate for particular kinds of gear. Gadgets with electric engines, for instance, may encounter diminished effectiveness and expanded heat age when fueled by a changed sine wave. Nonetheless, numerous home devices and instruments can work really with a changed sine wave power supply, settling on these inverters a reasonable decision for specific applications.

Notwithstanding waveform characterization, energy inverters can be arranged in view of their power limit. Inverter sizes are usually estimated in kilowatts (kW) or megawatts (MW) for bigger frameworks. Limited scope applications, for example, controlling a PC or a fridge during a blackout, may require inverters in the scope of two or three hundred watts. Private sunlight based establishments regularly use inverters with limits going from 1 kW to 10 kW, contingent upon the size of the framework.

Business and modern applications frequently request bigger inverters, and utility-scale sun based or wind ranches may convey inverters with limits in the megawatt range. The power limit of an inverter is a basic consider deciding the general framework effectiveness and the quantity of boards or sources it can oblige. Matching the inverter size to the particular necessities of the application is fundamental for ideal execution and cost-adequacy.

The combination of energy stockpiling frameworks into power frameworks has become progressively common, and this has prompted the improvement of inverter-chargers. Inverter-chargers are gadgets that join the elements of an inverter and a battery charger, empowering the consistent coordination of energy stockpiling into a power framework. These gadgets are regularly utilized in off-network and half breed frameworks.

In off-framework establishments, the inverter-charger guarantees a steady AC power supply while likewise charging the batteries when an outside power source, like a generator or sunlight powered chargers, is accessible. During times of low energy creation, the inverter draws power from the batteries to fulfill the heap need.

Mixture frameworks consolidate sustainable power sources with lattice availability and energy stockpiling, and the inverter-charger deals with the progression of energy between these sources, enhancing energy use in view of elements, for example, energy costs, matrix soundness, and ecological contemplations.

As innovation keeps on propelling, the scene of energy inverters is advancing with the rise of shrewd inverters. Shrewd inverters integrate correspondence and control abilities, empowering ongoing checking, controller, and lattice cooperation highlights. These inverters can speak with utility suppliers and lattice administrators to add to matrix security, answer framework signals, and improve the general proficiency and unwavering quality of the power framework.

One vital element of brilliant inverters is their capacity to give matrix support capabilities, like voltage and recurrence guideline. This ability becomes fundamental as the infiltration of sustainable power sources increments, bringing changeability and discontinuity into the lattice. Brilliant inverters effectively partake in framework the executives, offering auxiliary types of assistance and alleviating difficulties related with the fluctuating idea of environmentally friendly power age.

The idea of a "brilliant framework" is firmly connected to the organization of savvy inverters. A brilliant framework coordinates progressed correspondence and control innovations to upgrade the productivity, dependability, and maintainability of the electric power framework. Shrewd inverters assume an imperative part in this change by working with bidirectional correspondence between the utility and the dispersed energy assets, empowering a more powerful and responsive matrix foundation.

Energy versatility has turned into a noticeable thought in power framework configuration, stressing the significance of guaranteeing a dependable power supply despite catastrophic events, lattice disappointments, or different disturbances. Inverters, especially those incorporated with energy capacity frameworks, contribute essentially to improving energy strength.

Microgrid frameworks act as a commendable utilization of energy flexibility standards. A microgrid is a confined energy framework that can work freely or related to the fundamental matrix. It regularly comprises of circulated energy assets, like sun powered chargers, wind turbines, energy capacity, and high level inverters. During a network blackout, a microgrid can keep on providing capacity to basic burdens, guaranteeing that fundamental administrations stay functional.

High level inverters inside microgrids empower consistent change between lattice associated and islanded modes. Islanding alludes to the capacity of a microgrid to separate from the fundamental matrix and work independently.

In such situations, the inverters assume a significant part in keeping up with the harmony between energy age and utilization inside the microgrid, guaranteeing soundness and dependability during separation.

The multiplication of electric vehicles (EVs) plays additionally extended the part of energy inverters. Electric vehicle charging stations frequently integrate inverters to change over network power into the immediate flow (DC) expected for charging EV batteries. Moreover, bi-directional inverters empower vehicle-to-framework (V2G) capacities, permitting EVs to release put away energy back to the lattice during top interest periods. This bidirectional progression of energy between the matrix and EVs adds to network steadiness and gives a potential income stream to EV proprietors.

As innovation keeps on progressing, new materials and plans are being investigated to improve the proficiency and execution of energy inverters. Silicon carbide (SiC) and gallium nitride (GaN) are arising as elective semiconductor materials for power gadgets. These materials offer higher warm conductivity and lower exchanging misfortunes contrasted with customary silicon-based gadgets, bringing about additional productive and smaller inverters.

The quest for higher productivity and lower ecological effect has likewise prompted the advancement of cutting edge inverters with further developed power thickness. High-recurrence exchanging, high level cooling strategies, and inventive circuit geographies add to lessening the size and weight of inverters while keeping up with or working on their presentation. Reduced and lightweight inverters are especially important in applications where space and weight requirements are basic, like electric vehicles and compact power frameworks.

The coordination of computerized reasoning (simulated intelligence) and AI (ML) into energy the executives frameworks addresses one more boondocks in inverter innovation. Simulated intelligence driven calculations can enhance the activity of inverters by anticipating energy creation designs, load requests, and matrix conditions. This prescient ability empowers the inverter to proactively change its boundaries for ideal execution, adding to expanded productivity and framework strength.

In addition, simulated intelligence and ML can upgrade issue location and diagnostics in inverters. The capacity to recognize and resolve issues continuously further develops framework dependability and diminishes personal time. This proactive support approach is especially significant in remote or out of reach places where manual assessments might challenge.

3.2 Discuss the advantages and disadvantages of each type.

The investigation of different sorts of energy inverters uncovers an assorted scene, each with its arrangement of benefits and drawbacks. Understanding these qualities is vital for choosing the most reasonable inverter for a particular application. We should dig into the advantages and disadvantages of various sorts of energy inverters, including matrix tie inverters, off-network inverters, string inverters, microinverters, unadulterated sine wave inverters, changed sine wave inverters, and inverter-chargers.

Network tie inverters, intended for associating sustainable power frameworks to the electrical matrix, accompany their own arrangement of benefits and burdens. One remarkable benefit is their capacity to contribute overabundance energy back to the lattice, advancing manageability and possibly prompting cost reserve funds through net metering. Network attach inverters empower clients to use the current framework foundation, guaranteeing a dependable power supply in any event, when sustainable sources are deficient. Be that as it may, a critical disadvantage is their dependence on matrix network; assuming the lattice goes down, framework secure inverters normally shut to forestall backfeeding power into the network, representing a wellbeing danger for utility specialists.

Interestingly, off-network inverters are vital for independent power frameworks, giving autonomy from the utility matrix. The principal advantage lies in their capacity to work in distant places where matrix network is missing or questionable. Off-framework inverters empower clients to produce and store their own power, guaranteeing a solid energy supply in off-network applications like lodges, boats, and far off correspondence towers. Be that as it may, a significant disservice is the requirement for energy capacity arrangements, normally batteries, to store overabundance energy for use during times of low creation or popularity. The expense and upkeep of these batteries can be a critical disadvantage for off-network frameworks.

Inside the domain of network tie inverters, the qualification between string inverters and microinverters offers shifting benefits and impediments. String inverters are financially savvy and appropriate for enormous scope establishments because of their capacity to deal with different sun powered chargers wired in series. The disadvantage, nonetheless, is that the presentation of the whole string can be influenced by concealing or the underperformance of a solitary board. The presentation of Most extreme Power Point Following (MPPT) innovation has relieved this downside somewhat by streamlining the result of each board.

Microinverters, then again, offer individual advancement for each sun powered charger, limiting the effect of concealing or glitches on the general framework execution. This plan adaptability is favorable, especially in establishments where boards have fluctuating directions or are liable to concealing at various times. Notwithstanding, the expanded intricacy of microinverter frameworks can prompt higher forthright expenses, and the upkeep of individual inverters on each board might introduce difficulties concerning availability.

The waveform arrangement of energy inverters into unadulterated sine wave and changed sine wave inverters additionally accompanies particular benefits and burdens. Unadulterated sine wave inverters create a smooth and steady waveform that intently imitates the regular AC power given by the utility network. This top notch power yield makes them reasonable for a large number of uses, including delicate gadgets, engines, and machines. The burden, be that as it may, is the greater expense related with unadulterated sine wave inverters contrasted with their adjusted sine wave partners.

Changed sine wave inverters, while more prudent, produce a ventured waveform that may not be reasonable for particular kinds of gear. Gadgets with electric engines, for instance, may encounter diminished proficiency and expanded heat age when controlled by a changed sine wave. In any case, numerous home devices and instruments can work successfully with a changed sine wave power supply, making these inverters a reasonable and practical decision for specific applications.

As far as power limit, the benefits and inconveniences are intently attached to the particular necessities of the application. Limited scope applications, for example, driving a PC or a cooler during a blackout, may profit from more modest inverters with limits in the scope of a couple hundred watts. Then again, business and modern applications, as well as utility-scale sun powered or wind ranches, require bigger inverters with limits in the kilowatt or megawatt range. The upside of bigger inverters lies in their capacity to deal with higher loads and oblige a more noteworthy number of boards or energy sources. In any case, the disadvantage is the related higher forthright expense and possible over-measuring for more modest establishments.

The joining of energy stockpiling frameworks through inverter-chargers presents its own arrangement of benefits and burdens. Inverter-chargers are urgent for off-lattice and cross breed frameworks, offering consistent incorporation of energy stockpiling into the power framework. The benefit lies in their capacity to give a steady AC power supply while at the same time charging batteries when an outer power source is free. This guarantees energy flexibility and dependable power supply during times of low energy creation. Nonetheless, the inconvenience incorporates the expense of energy stockpiling arrangements, like batteries, and the requirement for standard support.

Brilliant inverters, furnished with correspondence and control abilities, offer a few benefits as far as framework backing and the executives. Their capacity to add to framework steadiness, answer lattice signals, and upgrade generally productivity makes them significant resources in the developing energy scene. The bidirectional correspondence between brilliant inverters and utility suppliers works with a more powerful and responsive lattice framework, adding to the idea of a shrewd network. In any case, the intricacy and extra highlights of shrewd inverters might prompt higher forthright expenses and possibly expanded upkeep necessities.

The use of energy strength standards in microgrid frameworks is another region where inverters assume a basic part. The upside of microgrids lies in their capacity to work independently during network blackouts, guaranteeing a nonstop power supply to basic burdens. High level inverters inside microgrids empower consistent changes between matrix associated and islanded modes, keeping up with security and unwavering quality during detachment. Nonetheless, the downside remembers the underlying venture for laying out a microgrid framework and the continuous upkeep necessities.

In the domain of electric vehicles (EVs), inverters utilized in charging stations and bidirectional inverters for vehicle-to-lattice (V2G) capacities offer benefits as far as network steadiness and potential income streams. Accusing stations prepared of inverters empower the change of framework power into DC power expected for charging EV batteries. Bidirectional inverters empower EVs to release put away energy back to the lattice during top interest periods, adding to framework solidness. Notwithstanding, challenges incorporate the requirement for broad charging framework, normalization of charging conventions, and potential worries in regards to the effect on EV battery duration.

The investigation of cutting edge materials, like silicon carbide (SiC) and gallium nitride (GaN), presents benefits regarding expanded productivity and execution of energy inverters. These materials offer higher warm conductivity and lower exchanging misfortunes contrasted with customary silicon-based gadgets. The benefits incorporate better productivity, decreased size and weight, and improved generally speaking execution. Be that as it may, the potential disadvantages might incorporate higher material expenses and the requirement for particular assembling processes.

The combination of man-made reasoning (simulated intelligence) and AI (ML) into energy the executives frameworks presents benefits as far as streamlining inverter activity and upgrading shortcoming recognition. Simulated intelligence driven calculations can anticipate energy creation designs, load requests, and network conditions, empowering proactive changes for ideal execution. Shortcoming recognition and diagnostics through man-made intelligence and ML add to further developed framework unwavering quality and decreased free time. Be that as it may, difficulties might incorporate the requirement for refined calculations, information security contemplations, and expected intricacies in execution.

All in all, each kind of energy inverter accompanies its extraordinary arrangement of benefits and impediments, making it critical to consider the particular prerequisites of the application cautiously. Lattice tie inverters offer the advantages of matrix availability and likely expense investment funds through net metering yet face impediments during framework blackouts. Off-framework inverters give freedom from the lattice yet require energy capacity arrangements, adding to the general expense. String inverters are savvy for enormous establishments yet may experience the ill effects of decreased execution because of concealing, while microinverters offer individual board enhancement at a higher forthright expense.

The waveform order presents a compromise between the excellent result of unadulterated sine wave inverters and the expense viability of changed sine wave inverters. Power limit contemplations rely upon the size of the application, with bigger inverters offering higher loads yet possibly bringing about greater expenses. Inverter-chargers work with energy capacity combination however accompany extra expenses and support prerequisites. Brilliant inverters add to matrix solidness however may have higher forthright expenses and expanded intricacy.

Microgrid frameworks offer energy flexibility however require starting venture and continuous support. Inverters in EV charging stations and bidirectional inverters for V2G capacities add to network security yet face difficulties connected with charging foundation and normalization. The investigation of cutting edge materials, for example, SiC and GaN, offers further developed proficiency and execution yet may involve greater expenses. Man-made intelligence and ML combination improve inverter activity yet may require complex calculations and contemplations for information security.

As innovation keeps on propelling, the continuous advancement of energy inverters will probably address a portion of these difficulties and push the limits of proficiency, dependability, and maintainability. Eventually, the determination of an energy inverter ought to be founded on a careful investigation of the particular necessities and requirements of the application, gauging the benefits and weaknesses to accomplish the most reasonable and viable arrangement.

3.3 Highlight real-world applications for different types of energy inverters.

The different scene of energy inverters finds down to earth applications across different certifiable situations, each type serving explicit necessities and prerequisites. By looking at the particular elements of various inverters, we can feature their genuine applications and comprehend how they add to various parts of our day to day routines and more extensive energy frameworks.

Framework tie inverters, intended to associate sustainable power frameworks to the electrical matrix, have far reaching applications in both private and business settings. In private sun oriented photovoltaic (PV) establishments, framework tie inverters assume a focal part in changing over DC power created by sunlight based chargers into AC power viable with the matrix. This permits property holders to balance their power utilization with spotless, sustainable power and possibly contribute overabundance power back to the network.

Business and modern offices likewise benefit from lattice tie inverters in bigger scope sun powered establishments. By coordinating sun based clusters with matrix tie inverters, organizations can outfit sun oriented energy to diminish their reliance on ordinary lattice power and lower power costs. Lattice tie inverters empower these frameworks to flawlessly interface with the matrix, permitting overabundance energy to be taken care of once more into the lattice and possibly prompting monetary impetuses through net metering game plans.

With regards to electric vehicle (EV) charging framework, matrix tie inverters are utilized to change over lattice power into the necessary organization for charging EV batteries. Accusing stations prepared of framework tie inverters work with the proficient and solid charging of electric vehicles, adding to the boundless reception of clean transportation arrangements.

On the other hand, off-framework inverters find application in situations where an association with the utility lattice is either inaccessible or temperamental. Distant areas, for example, off-lattice lodges, boats, or correspondence towers, frequently depend on off-framework inverters to change over DC power from sources like sunlight based chargers or batteries into AC power for day to day use. Off-matrix inverters are a foundation of independent power frameworks, guaranteeing a dependable and ceaseless power supply in regions without admittance to concentrated power framework.

Crisis reinforcement frameworks additionally regularly integrate off-lattice inverters. In case of a matrix blackout, these inverters permit basic burdens to be fueled by energy put away in batteries. Medical clinics, server farms, and different offices where continuous power is critical utilize off-framework inverters to keep up with fundamental tasks during network disturbances.

String inverters, with their savvy configuration appropriate for huge scope establishments, are common in both private and business sun powered projects. In private applications, string inverters are utilized to change over DC power created by a progression of interconnected sunlight based chargers into AC power for family use. While concealing or underperformance of one board might influence the whole string, headways in Most extreme Power Point Following (MPPT) innovation have relieved these worries, going with string inverters a pragmatic decision for some property holders.

In business sunlight based establishments, where the size of the framework is bigger, string inverters keep on being practical and effective. They handle different strings of sunlight based chargers, changing over the amassed DC power into AC power for dissemination. The versatility of string inverters makes them reasonable for different business applications, from driving places of business to huge scope modern buildings.

Microinverters, with their singular streamlining of each sunlight based charger, find applications in private establishments where board directions change, or concealing is a worry. By introducing microinverters straightforwardly on each board, these frameworks limit the effect of concealing or underperformance of individual boards on the general energy yield. This plan adaptability makes microinverters appropriate for private housetops with changing sun openness over the course of the day.

Microinverters are likewise significant in conditions where style and space usage are basic. For instance, neighborhoods with restricted rooftop space or building requirements benefit from the dispersed and minimized nature of microinverter frameworks.

Unadulterated sine wave inverters, known for creating great AC power, find applications in situations where the unwavering quality and accuracy of electrical result are fundamental. Numerous domestic devices, touchy hardware, and clinical gear work all the more proficiently and securely with an unadulterated sine wave power supply.

Clinical offices, labs, and research establishments frequently send un-adulterated sine wave inverters to guarantee the trustworthiness of gear and investigations. Also, in private settings, property holders might pick unadulterated sine wave inverters for reinforcement power frameworks to ensure the ideal exhibition of apparatuses and electronic gadgets.

Altered sine wave inverters, while more affordable, are reasonable for applications where the particular waveform necessities of gadgets are less basic. Numerous normal domestic devices and instruments can work really with a changed sine wave power supply, making these inverters commonsense for general use.

In sporting settings, for example, setting up camp or sailing, changed sine wave inverters power a scope of gadgets, including lights, coolers, and theater setups. While certain gadgets might show decreased profi-ciency or produce more intensity when fueled by a changed sine wave, the expense viability of these inverters goes with them a famous decision for off-lattice and convenient power arrangements.

Inverter-chargers, joining the elements of an inverter and a battery charger, have flexible applications in off-network and cross breed frame-works. Off-matrix homes and lodges use inverter-chargers to change over DC power from inexhaustible sources or generators into AC power for everyday use while at the same time charging energy stockpiling bat-teries. This double usefulness guarantees a steady power supply and the renewal of energy stockpiling during times of adequate power age.

Half and half frameworks that coordinate environmentally friendly power sources with lattice network and energy stockpiling benefit from the abilities of inverter-chargers. These frameworks wisely deal with the progression of energy, advancing the utilization of inexhaustible sources, drawing power from the lattice when required, and putting away abundance energy for sometime in the future. Inverter-chargers assume an essential part in adjusting energy age and utilization, adding to the general proficiency of mixture power frameworks.

Savvy inverters, outfitted with correspondence and control abili-ties, are at the very front of network modernization endeavors. These inverters add to true applications pointed toward improving framework strength, supporting environmentally friendly power incorporation, and empowering progressed lattice the board.

One striking application is the arrangement of network support capabilities, like voltage and recurrence guideline. Savvy inverters effectively partake in balancing out the framework by changing their result in light of lattice conditions. This ability turns out to be progressively essential as the extent of environmentally friendly power sources on the lattice develops, presenting inconstancy and irregularity.

Microgrid frameworks, intended to work autonomously or related to the primary network, influence the abilities of shrewd inverters. In case of a matrix blackout, brilliant inverters inside microgrids empower a consistent change to islanded mode, guaranteeing ceaseless power supply to basic burdens. Microgrid frameworks track down applications in different settings, including army bases, far off networks, and modern offices, where energy strength and dependability are principal.

The arrangement of energy stockpiling frameworks, coordinated with inverter-chargers and shrewd inverters, adds to certifiable applications in improving matrix unwavering quality and soundness. Energy capacity matched with shrewd inverters considers the effective administration of irregular sustainable power sources, empowering the capacity of abundance energy during times of high creation and its delivery during seasons of popularity or low creation.

In the domain of electric vehicles (EVs), bidirectional inverters assume an essential part in certifiable applications like vehicle-to-matrix (V2G) innovation. These inverters empower EVs to release put away energy back into the matrix during top interest periods, giving a potential income stream to EV proprietors and adding to framework soundness. V2G innovation holds guarantee in adjusting matrix stacks and upgrading the use of environmentally friendly power.

High level materials, including silicon carbide (SiC) and gallium nitride (GaN), are tracking down applications in energy inverters to improve effectiveness and execution. SiC and GaN-based inverters are utilized in electric vehicles, environmentally friendly power frameworks, and other high-power applications. Their unrivaled warm conductivity and lower exchanging misfortunes add to expanded proficiency and decreased size, tending to the interest for conservative and proficient influence transformation arrangements.

The coordination of man-made consciousness (artificial intelligence) and AI (ML) into energy the executives frameworks, combined with cutting edge inverters, is changing this present reality scene of energy frameworks. Man-made intelligence driven calculations advance the activity of inverters by anticipating energy creation designs, load requests, and framework conditions. This prescient ability permits inverters to proactively change their boundaries for ideal execution, adding to expanded proficiency and lattice steadiness.

Man-made intelligence and ML applications in issue identification and diagnostics improve the unwavering quality of energy inverters in true settings. The capacity to distinguish and resolve issues progressively diminishes personal time and works on the general execution of force frameworks.

4

Chapter 4

Green Energy and Inverters

Efficient power energy and inverters assume crucial parts in the continuous worldwide work to progress towards a more maintainable and harmless to the ecosystem energy scene. As the world wrestles with the difficulties presented by environmental change and the consumption of customary petroleum derivative assets, the journey for cleaner and sustainable power sources has picked up uncommon speed. This change in perspective towards efficient power energy is set apart by a large number of innovations and developments, among which inverters stand apart as vital parts in outfitting and streamlining the capability of environmentally friendly power frameworks.

Environmentally friendly power sources, for example, sun oriented, wind, hydro, and geothermal, are at the very front of the environmentally friendly power energy upset. These sources offer the commitment of cleaner power age with altogether lower natural effects contrasted with customary petroleum products. Sun oriented energy, specifically, has seen a surprising flood in fame and reception. Photovoltaic (PV) innovation, which changes over daylight into power, has turned into a foundation of efficient power energy drives around the world.

Sun oriented inverters are key in sun based power frameworks, filling in as the key part that changes over the immediate current (DC) produced by sun powered chargers into substituting current (AC), which is appropriate for use in homes and organizations. This change is essential on the grounds that most apparatuses and electrical gadgets work on AC power. Inverters guarantee that the power delivered by sunlight based chargers is viable with the network and can be flawlessly incorporated into existing power framework.

One of the essential benefits of sunlight based inverters lies in their capacity to work with net metering. Net metering permits sunlight based charger proprietors to take care of overabundance power once again into the framework, procuring credits that can be utilized to balance their energy utilization during periods when sun oriented creation is deficient. This unique cooperation with the network features the interconnected idea of sustainable power frameworks and conventional power matrices, underscoring the requirement for refined and effective inverters.

Notwithstanding sunlight based energy, wind power addresses one more essential part of the environmentally friendly power energy scene. Wind turbines saddle the motor energy of the breeze to produce power, and inverters assume a basic part in streamlining the exhibition of these frameworks. Similarly as in sunlight based power applications, wind inverters convert the variable and frequently capricious result of wind turbines into a stable and lattice viable AC yield.

The changeability of environmentally friendly power sources represents a one of a kind arrangement of difficulties that inverters should address. Dissimilar to customary power establishes that can give a consistent and unsurprising result, sun oriented and wind energy creation is dependent upon weather patterns and other natural elements. Inverters furnished with cutting edge control calculations and lattice the executives capacities assist with moderating the discontinuity of environmentally friendly power sources, guaranteeing a predictable and dependable power supply.

As the interest for environmentally friendly power keeps on rising, the significance of network tied inverters turns out to be progressively apparent. Matrix tied inverters, otherwise called lattice associated inverters, empower environmentally friendly power frameworks to communicate consistently with the electrical network. This joining works with the infusion of overabundance power into the lattice during times of high environmentally friendly power creation and considers the withdrawal of power from the network when inexhaustible sources are lacking.

Microinverters address an eminent development in the domain of sun oriented power. Not at all like conventional string inverters that are associated with various sunlight powered chargers in series, microinverters are introduced on individual sunlight based chargers.

This decentralized methodology offers a few benefits, including expanded energy reap, further developed framework dependability, and improved checking capacities. Microinverters additionally alleviate the effect of halfway concealing on sunlight powered chargers, considering ideal energy creation even in testing ecological circumstances.

With regards to environmentally friendly power energy, the job of inverters stretches out past sun oriented and wind applications. Mixture inverters, for example, are intended to work with different wellsprings of energy, frequently consolidating sunlight based and battery capacity frameworks. This flexibility empowers clients to boost the use of sustainable power and store abundance power for sometime in the future, tending to the innate discontinuity of sun based and wind power.

Battery capacity frameworks, combined with inverters, assume a crucial part in accomplishing energy freedom and strength. Energy capacity takes into account the catch and maintenance of overabundance energy created during times of high environmentally friendly power creation. This put away energy can then be sent during times when sustainable sources are not effectively producing power, giving a ceaseless and solid power supply. Inverters in these frameworks deal with the charging and releasing of batteries, advancing the general proficiency and execution of the energy stockpiling arrangement.

The coordination of inverters with energy capacity frameworks likewise opens up open doors for off-matrix applications. In far off regions or during crises, where admittance to the customary power framework is restricted or inaccessible, off-matrix inverters empower the making of independent power frameworks. These frameworks regularly consolidate environmentally friendly power sources, like sunlight based chargers and wind turbines, alongside energy capacity answers for give a self-supporting wellspring of power.

The coming of brilliant matrix advances further highlights the meaning of inverters in the developing energy scene. Brilliant inverters, furnished with cutting edge correspondence capacities, empower bidirectional data stream between the utility network and individual power frameworks. This bidirectional correspondence works with ongoing checking, control, and enhancement of conveyed energy assets, adding to a stronger and responsive matrix foundation.

The change to efficient power energy isn't just determined by natural contemplations yet in addition by financial variables. The expense intensity of environmentally friendly power innovations, combined with the declining costs of sunlight based chargers and wind turbines, has made efficient power energy progressively reasonable according to a monetary viewpoint. Inverters, as fundamental parts of environmentally friendly power frameworks, add to the by and large monetary practicality of efficient power energy projects by guaranteeing productive energy change and matrix combination.

Chasing manageable energy arrangements, perceiving the significance of strategy systems and administrative support is fundamental. States all over the planet assume a pivotal part in boosting the reception of efficient power energy innovations and establishing a climate helpful for sustainable power improvement. Motivators, for example, feed-in duties, tax breaks, and sustainable portfolio principles urge organizations and people to put resources into environmentally friendly power frameworks, driving the interest for inverters and related advancements.

The worldwide work to battle environmental change has prompted aggressive focuses for decreasing ozone harming substance emanations, with numerous nations focusing on accomplishing net-no outflows in the next few decades. In this specific situation, the job of inverters grows past their specialized capabilities to become key empowering influences of the change to a low-carbon economy. The charge of different areas, including transportation and warming, depends on the accessibility of perfect and environmentally friendly power sources, with inverters assuming a focal part in guaranteeing the consistent coordination of these sources into the more extensive energy biological system.

As the environmentally friendly power energy upheaval picks up speed, innovative work endeavors are centered around improving the proficiency and execution of inverters. Enhancements in power gadgets, control calculations, and materials are driving the development of inverter advances. High-proficiency inverters not just add to the general adequacy of sustainable power frameworks yet additionally straightforwardly affect decreasing energy misfortunes and expanding the financial feasibility of efficient power energy projects.

The idea of a decentralized and democratized energy framework is a focal principle of the environmentally friendly power energy change. Circulated energy assets, remembering limited scope sun based establishments for private housetops, local area sun oriented projects, and microgrids, typify this decentralized methodology. Inverters, especially microinverters and shrewd inverters, assume a vital part in empowering the consistent mix of these conveyed assets into the more extensive energy framework.

The meaning of environmentally friendly power energy and inverters is additionally highlighted by their capability to address energy access difficulties in creating locales. In many regions of the planet, particularly in remote and off-matrix regions, admittance to dependable power stays a critical obstacle. Off-network environmentally friendly power frameworks, fueled by sunlight based chargers and upheld by inverters, offer a practical and versatile answer for carry power to underserved networks. This not just works on the personal satisfaction for people here yet additionally adds to monetary turn of events and social strengthening.

In the mission for an economical future, the significance of schooling and mindfulness couldn't possibly be more significant. Public mindfulness and comprehension of efficient power energy innovations, including the job of inverters, are fundamental for encouraging a culture of

manageability. Instructive drives, local area outreach projects, and associations between the scholarly world, industry, and government organizations are instrumental in advancing the boundless reception of efficient power energy arrangements.

Difficulties and obstructions continue in the broad reception of environmentally friendly power energy and inverters. One huge test is the discontinuity and changeability of sustainable power sources, which can prompt vacillations in power yield. Inverters, outfitted with cutting edge control and energy the executives highlights, assist with relieving these difficulties by giving lattice security and guaranteeing a smooth change among sustainable and ordinary power sources.

One more test lies in the combination of sustainable power frameworks into existing foundations. The fluctuation in energy creation from sources like sun oriented and wind requires strong lattice the executives techniques. Savvy inverters, with their capacity to speak with the matrix and answer ongoing circumstances, assume a urgent part in tending to these combination challenges. Notwithstanding, refreshing and modernizing the current framework foundation stay key needs for guaranteeing the consistent reconciliation of environmentally friendly power energy advances.

The natural effect of assembling and discarding sun powered chargers and other sustainable power parts is a thought that goes with the environmentally friendly power energy progress. Endeavors to foster feasible assembling practices and reusing answers for end-of-life parts are fundamental to limit the biological impression of environmentally friendly power advances. Inverters, as vital parts of these frameworks, are likewise dependent upon these contemplations, featuring the requirement for a comprehensive way to deal with maintainability in the whole lifecycle of efficient power energy foundation.

The advancing scene of energy stockpiling innovations presents the two amazing open doors and difficulties for inverters. While energy capacity arrangements upgrade the unwavering quality and adaptability of sustainable power frameworks, the materials and advances utilized in batteries present natural worries. Inverters, related to progressions in battery innovation and reusing strategies, can add to the improvement of feasible energy stockpiling arrangements.

Online protection arises as a basic part of the cutting edge energy framework, including efficient power energy frameworks and inverters. As these frameworks become progressively associated and dependent on advanced innovations, they become possible focuses for digital dangers. Guaranteeing the online protection of inverters and environmentally friendly power framework is basic to shielding the dependability and honesty of the general energy biological system.

4.1 Examine the integration of energy inverters with renewable energy sources (solar, wind, etc.).

The incorporation of energy inverters with sustainable power sources addresses a vital part of the worldwide change towards a more feasible and harmless to the ecosystem energy scene. Environmentally friendly power sources, for example, sunlight based and wind, have arisen as central participants in the mission to diminish fossil fuel byproducts and moderate the effect of environmental change. In this specific situation, energy inverters assume a basic part in enhancing the proficiency and unwavering quality of environmentally friendly power frameworks, guaranteeing the consistent joining of clean power into the current energy foundation.

Sun powered energy, saddled through photovoltaic (PV) innovation, has seen a striking flood in reception as a suitable and versatile environmentally friendly power source. Sun powered chargers create direct flow (DC) power when presented to daylight. Be that as it may, the electrical frameworks in homes and organizations commonly work on exchanging flow (AC). This is where inverters become possibly the most important factor. Sun powered inverters are central parts that convert the DC yield from sunlight powered chargers into AC, making it viable with the network and usable for different applications.

The essential capability of sun based inverters is to work with the productive change of sun based power, empowering its reconciliation into the power matrix or for direct use in homes and organizations. In doing as such, they add to decreasing reliance on traditional petroleum derivatives and alleviating the natural effect related with power age. The capacity to consistently interface sun based power frameworks to the lattice is fundamental for expanding the advantages of sun oriented energy, and inverters act as the vital empowering influences of this joining.

Besides, sun oriented inverters assume a pivotal part in empowering net metering, a component that permits sun powered charger proprietors to take care of overabundance power once more into the matrix. During times of high sunlight based energy creation, when the power produced surpasses the prompt interest, the excess power can be sent back to the lattice. This not just aides in balancing the energy utilization from the network during times of low sunlight based creation yet additionally gives a road to sun powered charger proprietors to procure credits or pay for the power they add to the matrix.

The incorporation of sun oriented inverters isn't restricted to huge scope sun powered ranches; it reaches out to private and business applications too. In the private area, housetop sun based establishments are turning out to be progressively normal, and inverters customized for these applications assume a basic part in streamlining energy creation at the singular level. The decentralization of sunlight based power age, worked with by private sun powered chargers and inverters, adds to a more disseminated and strong energy foundation.

Wind power is one more huge player in the sustainable power scene, and inverters assume a vital part in streamlining the exhibition of wind energy frameworks. Wind turbines convert the motor energy of the breeze into electrical power, commonly creating power as AC. Notwithstanding, the result of wind turbines can be variable and eccentric, contingent upon elements, for example, wind speed and course. Wind inverters address these difficulties by changing over the variable AC result of wind turbines into a stable and matrix viable structure.

Similarly as in sun oriented applications, the joining of wind inverters is fundamental for the consistent connection between wind power frameworks and the electrical lattice. Framework tied inverters, intended for wind energy applications, empower the infusion of wind-created power into the matrix during times of high creation. This framework connection works with the adjusting of power organic market, adding to network solidness and unwavering quality.

The inconstancy of sustainable power sources represents a remarkable arrangement of difficulties that inverters should address. Dissimilar to conventional power establishes that give a consistent and unsurprising result, sun oriented and wind energy creation varies in view of weather patterns and other ecological elements. Inverters furnished with cutting edge control calculations and framework the board capacities assist with moderating the discontinuity of sustainable power sources. These savvy inverters effectively deal with the progression of power, guaranteeing a steady and dependable power supply even notwithstanding factor environmentally friendly power age.

Microinverters address an outstanding development in the domain of sunlight based power mix. Dissimilar to conventional string inverters, which are associated with various sunlight based chargers in series, microinverters are introduced on individual sun powered chargers. This decentralized methodology offers a few benefits. Microinverters augment energy collect by freely improving the presentation of each sun powered charger. This differentiations with string inverters, where the exhibition of one board can affect the result of the whole string.

The decentralized idea of microinverters improves framework dependability. If one sun powered charger or microinverter glitches, it doesn't influence the whole framework's presentation, just like with string inverters. Moreover, microinverters alleviate the effect of halfway concealing on sunlight based chargers. In conventional frameworks, concealing on one board could altogether diminish the result of the whole string. Microinverters guarantee that the concealed board's effect is restricted to itself, permitting the remainder of the framework to ideally work.

Crossover inverters address one more feature of the joining of sustainable power sources, consolidating numerous wellsprings of energy, frequently sunlight based and battery capacity frameworks. These inverters are intended to work flawlessly with both environmentally friendly power information sources and energy stockpiling frameworks, offering a flexible answer for upgrading energy use.

Mixture frameworks permit clients to boost the usage of sustainable power, store overabundance power for sometime in the future, and give a persistent power supply in any event, when inexhaustible sources are not effectively creating power.

Battery capacity frameworks, combined with inverters, assume a significant part in accomplishing energy freedom and versatility. These frameworks store overabundance energy created during times of high environmentally friendly power creation for use during times when sustainable sources are inadequate. Inverters in these frameworks deal with the charging and releasing of batteries, upgrading the general effectiveness and execution of the energy stockpiling arrangement.

The incorporation of inverters with energy capacity frameworks opens up open doors for off-matrix applications. In distant regions or during crises, where admittance to the conventional power matrix is restricted or inaccessible, off-network inverters empower the production of independent power frameworks. These frameworks normally consolidate environmentally friendly power sources, like sunlight based chargers and wind turbines, alongside energy capacity answers for give a self-supporting wellspring of power.

The idea of a decentralized and democratized energy framework is a focal principle of the efficient power energy progress. Appropriated energy assets, remembering limited scope sun oriented establishments for private roofs, local area sun powered projects, and microgrids, embody this decentralized methodology. Inverters, especially microinverters and shrewd inverters, assume a significant part in empowering the consistent combination of these circulated assets into the more extensive energy framework.

Network tied inverters are major parts with regards to circulated energy assets. These inverters empower the association of limited scope environmentally friendly power frameworks to the matrix, permitting overabundance power to be taken care of once again into the network. This lattice tied setup empowers people, networks, and organizations to take part in the age and conveyance of clean energy effectively. Network tied inverters guarantee that the power created locally can be imparted to the more extensive local area, adding to a stronger and interconnected energy lattice.

The coming of brilliant framework advancements further highlights the meaning of inverters in the developing energy scene. Savvy inverters, outfitted with cutting edge correspondence capacities, empower bidirectional data stream between the utility matrix and individual power frameworks. This bidirectional correspondence works with constant observing, control, and streamlining of circulated energy assets, adding to a stronger and responsive matrix framework.

Brilliant inverters assume a urgent part in framework the board by giving important experiences into the presentation of disseminated energy assets. They empower utilities to screen and deal with the progression of power continuously, expect likely issues, and advance the general

proficiency of the framework. The two-way correspondence between shrewd inverters and the framework takes into account dynamic changes in accordance with voltage and recurrence, upgrading the solidness and unwavering quality of the electrical foundation.

As the interest for environmentally friendly power keeps on rising, the significance of framework tied inverters turns out to be progressively clear. Matrix tied inverters not just work with the coordination of sustainable power into the lattice yet in addition add to the security and dependability of the by and large electrical framework. The bidirectional progression of power between conveyed energy assets and the matrix features the interconnected idea of current energy frameworks and underlines the requirement for refined and productive inverters.

The change to efficient power energy isn't just determined by natural contemplations yet additionally by monetary elements. The expense seriousness of environmentally friendly power innovations, combined with the declining costs of sunlight based chargers and wind turbines, has made efficient power energy progressively reasonable according to a monetary point of view. Inverters, as fundamental parts of sustainable power frameworks, add to the generally monetary possibility of environmentally friendly power energy projects by guaranteeing productive energy transformation and network reconciliation.

The financial reasonability of environmentally friendly power projects is additionally improved by impetuses and strategy structures that advance their reception. States all over the planet assume an essential part in boosting the sending of environmentally friendly power advancements and establishing a climate helpful for supportable energy improvement. Motivations, for example, feed-in duties, tax reductions, and sustainable portfolio guidelines urge organizations and people to put resources into environmentally friendly power frameworks, subsequently driving the interest for inverters and related advancements.

The worldwide work to battle environmental change has prompted aggressive focuses for diminishing ozone depleting substance discharges, with numerous nations focusing on accomplishing net-no outflows in the next few decades. In this specific circumstance, the job of inverters grows past their specialized capabilities to become key empowering agents of the progress to a low-carbon economy. The charge of different areas, including transportation and warming, depends on the accessibility of spotless and sustainable power sources, with inverters assuming a focal part in guaranteeing the consistent reconciliation of these sources into the more extensive energy biological system.

The meaning of environmentally friendly power energy and inverters is additionally highlighted by their capability to address energy access difficulties in creating areas. In many regions of the planet, particularly in remote and off-network regions, admittance to solid power stays a huge obstacle.

Off-lattice sustainable power frameworks, controlled by sunlight based chargers and upheld by inverters, offer a supportable and versatile answer for carry power to underserved networks. This not just works on the personal satisfaction for people here yet in addition adds to financial turn of events and social strengthening.

Chasing manageable energy arrangements, perceiving the significance of instruction and awareness is fundamental. Public mindfulness and comprehension of efficient power energy innovations, including the job of inverters, are urgent for cultivating a culture of supportability. Instructive drives, local area outreach projects, and associations between the scholarly community, industry, and government organizations are instrumental in advancing the broad reception of efficient power energy arrangements.

Difficulties and hindrances persevere in the far and wide reception of environmentally friendly power energy and inverters. One huge test is the irregularity and changeability of environmentally friendly power sources, which can prompt vacillations in power yield. Inverters, furnished with cutting edge control and energy the board highlights, assist with moderating these difficulties by giving framework steadiness and guaranteeing a smooth progress among sustainable and customary power sources.

One more test lies in the combination of sustainable power frameworks into existing foundations. The fluctuation in energy creation from sources like sun oriented and wind requires strong lattice the executives techniques. Savvy inverters, with their capacity to speak with the matrix and answer constant circumstances, assume a pivotal part in tending to these mix difficulties. Nonetheless, refreshing and modernizing the current matrix foundation stay key needs for guaranteeing the consistent coordination of efficient power energy advances.

The natural effect of assembling and discarding sunlight powered chargers and other environmentally friendly power parts is a thought that goes with the efficient power energy progress. Endeavors to foster reasonable assembling practices and reusing answers for end-of-life parts are fundamental to limit the biological impression of sustainable power innovations. Inverters, as necessary parts of these frameworks, are likewise dependent upon these contemplations, featuring the requirement for an all encompassing way to deal with manageability in the whole lifecycle of efficient power energy foundation.

The developing scene of energy stockpiling advances presents the two valuable open doors and difficulties for inverters. While energy capacity arrangements upgrade the dependability and adaptability of sustainable power frameworks, the materials and advancements utilized in batteries present natural worries. Inverters, related to headways in battery innovation and reusing techniques, can add to the improvement of manageable energy stockpiling arrangements.

Network protection arises as a basic part of the cutting edge energy framework, including environmentally friendly power energy frameworks and inverters. As these frameworks become progressively associated and dependent on computerized innovations, they become expected focuses for digital dangers. Guaranteeing the network protection of inverters and environmentally friendly power foundation is basic to shielding the unwavering quality and trustworthiness of the general energy biological system.

4.2 Showcasing successful implementations of green energy systems with inverters.

Effective executions of environmentally friendly power energy frameworks with inverters act as unmistakable and rousing instances of the extraordinary effect of environmentally friendly power on different scales. From private sun powered establishments to enormous scope sun based and wind cultivates, these executions grandstand the flexibility and viability of environmentally friendly power energy advancements, with inverters assuming a focal part in enhancing energy creation, network combination, and by and large framework execution.

In the domain of private environmentally friendly power energy reception, housetop sun based establishments stand apart as a generally embraced and fruitful application. Mortgage holders overall are progressively going to sun powered chargers combined with inverters to tackle perfect and environmentally friendly power nearby. The effortlessness and versatility of private planetary groups make them open to an expansive scope of customers.

Inverter innovation is central to the outcome of private sunlight based establishments. String inverters, generally utilized in these applications, are associated with various sun powered chargers in series. They productively convert the immediate flow (DC) created by sunlight powered chargers into rotating flow (AC), making it viable with the electrical frameworks in homes. Thusly, these inverters guarantee that the sunlight based produced power can be flawlessly coordinated into the family's power supply.

Microinverters address one more fruitful execution in the private area. As opposed to string inverters, microinverters are introduced on individual sun powered chargers. This decentralized methodology improves energy creation by freely dealing with the result of each board. Microinverters alleviate the effect of concealing or execution varieties among boards, prompting expanded generally framework productivity.

Private efficient power energy frameworks frequently integrate net metering, an instrument empowered by inverters. Net metering permits mortgage holders to take care of overabundance power created by their sunlight based chargers back into the framework, procuring credits that can be utilized during times of low sun oriented creation. This advantageous connection between private sun based establishments, inverters, and the network features the interconnected idea of environmentally friendly power energy frameworks.

For a bigger scope, utility-scale sunlight based ranches grandstand the fruitful reconciliation of environmentally friendly power energy into the more extensive electrical lattice. These sun powered ranches, furnished with cutting edge inverters, produce critical measures of power from daylight. Matrix tied inverters assume an essential part in working with the infusion of sun powered created power into the lattice during top creation times.

The progress of utility-scale sunlight based ranches isn't just estimated by their ability to create clean energy yet in addition by their capacity to add to matrix steadiness. Inverters furnished with brilliant framework abilities empower consistent correspondence and coordination with the more extensive electrical foundation. This guarantees that the fluctuation inborn in sunlight based energy creation is overseen really, limiting disturbances to the framework.

Wind energy projects give one more convincing grandstand of effective environmentally friendly power energy executions, frequently in a joint effort with cutting edge inverters. Wind ranches, comprising of various breeze turbines, tackle the motor energy of the breeze to produce power. Inverters in these applications assume a basic part in changing over the variable substituting flow (AC) result of wind turbines into stable and network viable power.

Seaward wind ranches, arranged in waterways to catch solid and predictable breezes, embody the scale and aspiration of current breeze energy projects. Inverters intended for seaward wind applications are designed to endure unforgiving natural circumstances while guaranteeing solid and proficient energy change. These ventures contribute significantly to the environmentally friendly power blend, stressing the versatility and flexibility of wind energy frameworks with cutting edge inverters.

Crossover sustainable power frameworks, consolidating sun oriented, wind, and frequently energy capacity, epitomize the flexibility of efficient power energy executions. Inverters customized for half and half frameworks assume a critical part in dealing with the different wellsprings of energy. These frameworks upgrade the utilization of sun oriented and wind assets, store overabundance energy in batteries, and cleverly appropriate power in view of interest and matrix conditions.

The outcome of half breed frameworks is especially apparent in off-lattice and far off regions. These frameworks give a dependable and supportable wellspring of power where conventional matrix foundation is inaccessible. Inverters, working pair with energy capacity arrangements, guarantee a consistent power supply, tending to energy access difficulties and encouraging financial improvement in underserved networks.

Energy capacity frameworks coordinated with inverters address a groundbreaking answer for improving the dependability and versatility of efficient power energy executions.

Batteries, combined with cutting edge inverters, store abundance energy during times of high creation and delivery it when inexhaustible sources are lacking. This capacity tends to the irregularity of sustainable power, making it a more trustworthy and dispatchable asset.

Effective executions of efficient power energy frameworks with inverters reach out past power age. The jolt of transportation, worked with by inverters in electric vehicle (EV) charging framework, features the widening effect of efficient power energy advancements. Inverters deal with the progression of power from the lattice to charging stations, enhancing charging rate and effectiveness.

Brilliant urban communities are arising as center points for effective environmentally friendly power energy executions, utilizing trend setting innovations, including inverters, to upgrade manageability. These urban areas consolidate sustainable power sources, energy-proficient structures, and keen framework frameworks. Inverters furnished with correspondence capacities empower ongoing checking, control, and improvement of dispersed energy assets, adding to a stronger and responsive metropolitan foundation.

Microgrids, limited scope, confined energy frameworks with their own age and capacity abilities, embody the decentralization and independence reachable with efficient power energy executions. Inverters assume a focal part in microgrid control, guaranteeing consistent changes between sustainable power sources, stockpiling, and association with the primary network. This adaptability improves energy security and versatility, making microgrids effective answers for different applications, including distant networks and basic framework.

The outcome of environmentally friendly power energy executions is unpredictably connected to steady strategy systems and monetary motivations. States all over the planet assume a significant part in encouraging the reception of sustainable power innovations through strategies, for example, feed-in levies, tax reductions, and sustainable power principles. These actions establish a climate helpful for interest in efficient power energy projects, driving advancement and versatility.

Corporate drives and responsibilities to maintainability further add to the outcome of efficient power energy executions. Many organizations embrace environmentally friendly power as an essential part of their tasks, putting resources into sun based and wind ventures to diminish carbon impressions and meet supportability objectives. Inverters, with their job in improving energy creation and framework combination, have a critical impact in the progress of these corporate environmentally friendly power tries.

The financial suitability of environmentally friendly power energy executions is progressively obvious as the expenses of inexhaustible advancements, including sunlight based chargers and wind turbines, keep on declining. Inverters add to the monetary possibility of these undertakings by guaranteeing effective energy transformation and framework similarity.

The declining expenses of environmentally friendly power energy advances, combined with positive arrangement conditions, drive the far and wide reception of environmentally friendly power frameworks with inverters.

Local area based sustainable power projects represent the cooperative and comprehensive nature of effective environmentally friendly power energy executions. Local area sun powered projects, where various members put resources into a common sun oriented establishment, democratize admittance to sustainable power. Inverters in these ventures work with the coordination of individual sunlight based commitments into the more extensive electrical lattice, permitting members to benefit all in all from clean energy age.

Examples of overcoming adversity in efficient power energy executions frequently include a promise to schooling and mindfulness. Public commitment and comprehension of sustainable power advances, as well as the job of inverters, are fundamental for building backing and driving reception. Instructive projects, local area effort, and associations between instructive foundations and industry partners add to the achievement and manageability of efficient power energy drives.

Regardless of the progress of environmentally friendly power energy executions, difficulties and boundaries endure. The irregularity and fluctuation of environmentally friendly power sources, combined with the requirement for lattice modernization, present continuous difficulties. Inverters outfitted with cutting edge control and correspondence highlights address these difficulties by improving framework soundness and dealing with the powerful idea of sustainable power age.

The natural effect of assembling and discarding environmentally friendly power parts, including inverters, requires economical practices. Endeavors to create eco-accommodating assembling cycles and reusing arrangements are significant for limiting the biological impression of environmentally friendly power energy advances. Inverters, as vital parts of environmentally friendly power frameworks, are important for this more extensive manageability basic.

4.3 Discuss the environmental impact and sustainability aspects.

The conversation of ecological effect and manageability perspectives is key to grasping the more extensive ramifications of efficient power energy frameworks, including their related inverters. As the world wrestles with the results of environmental change and looks to progress away from conventional petroleum derivative based energy sources, a basic assessment of the ecological ramifications of sustainable power innovations becomes basic.

One of the essential inspirations driving the shift towards efficient power energy is the craving to alleviate the ecological effect related with traditional energy sources, especially the consuming of petroleum products.

Customary energy creation techniques, for example, coal and petroleum gas power plants, discharge huge measures of ozone harming substances, including carbon dioxide (CO_2), into the air. These discharges add to the nursery impact, prompting an Earth-wide temperature boost and environmental change.

Interestingly, sustainable power sources, for example, sun based and wind, produce power without radiating ozone harming substances during activity. Sun powered chargers saddle the sun's energy through photovoltaic cells, while wind turbines convert the active energy of the breeze into power. The organization of efficient power energy innovations, worked with by inverters, diminishes dependence on petroleum derivatives and adds to bringing down in general ozone harming substance discharges, consequently assisting with tending to the underlying drivers of environmental change.

While the functional period of efficient power energy frameworks shows clear natural advantages, the assembling, arrangement, and end-of-life stages likewise warrant examination. The development of sun powered chargers and wind turbines includes the extraction and handling of natural substances, energy-concentrated assembling cycles, and transportation. Likewise, the assembling of inverters requires materials like semiconductors, capacitors, and other electronic parts.

Endeavors are in progress to work on the manageability of assembling processes in the sustainable power area. The improvement of cleaner creation techniques, reusing drives, and the utilization of eco-accommodating materials are key systems to limit the natural effect of assembling sustainable power parts, including inverters. Developments in green assembling rehearses mean to decrease asset utilization, energy use, and discharges related with the creation of environmentally friendly power energy advancements.

Lifecycle evaluations (LCAs) give an extensive way to deal with assessing the natural effect of sustainable power frameworks, representing factors from unrefined substance extraction to end-of-life removal. LCAs assist with recognizing regions for development and guide the advancement of additional reasonable practices in the environmentally friendly power industry. Inverters, being essential parts of efficient power energy frameworks, are remembered for these evaluations to guarantee an all encompassing comprehension of their ecological impression.

The organization of sustainable power frameworks likewise includes land use contemplations. Enormous scope sunlight based and wind ranches, fundamental for satisfying the developing need for clean energy, require critical land regions. While these undertakings add to a cleaner energy blend, they can influence biological systems, biodiversity, and nearby networks. Key preparation and ecological effect evaluations are vital to adjust the advantages of sustainable power organization with the safeguarding of regular living spaces and local area prosperity.

Moreover, environmentally friendly power energy projects, including those utilizing inverters, should explore potential struggles connected with land use and biological system interruption. Adjusting the requirement for sustainable power improvement with ecological preservation is a continuous test, underscoring the significance of mindful siting, land the board, and local area commitment in efficient power energy drives.

The supportability of efficient power energy frameworks stretches out past their nearby ecological effect on contemplations of asset accessibility. The materials utilized in the assembling of sunlight powered chargers, wind turbines, and inverters incorporate uncommon earth components and other basic minerals. Guaranteeing a steady and practical inventory network for these materials is fundamental for the drawn out feasibility of environmentally friendly power innovations.

Endeavors to address material shortage include investigating elective materials, further developing reusing advances, and advancing roundabout economy standards. Reusing drives for sun powered chargers and wind turbines, including their related inverters, expect to recuperate significant materials and diminish the natural effect of extricating new assets. Economical asset the board techniques add to the versatility and life span of environmentally friendly power energy frameworks.

The finish of-life period of environmentally friendly power parts is one more viewpoint that requires consideration for accomplishing manageability. Sunlight powered chargers, wind turbines, and inverters have limited life expectancies, normally going from 20 to 30 years. Mindful removal and reusing rehearses are fundamental to forestall ecological pollution and guarantee the recuperation of important materials.

Reusing programs for sunlight powered chargers and wind turbines are developing to address the difficulties related with the removal of electronic waste. Developments in reusing advances empower the recuperation of metals, glass, and different materials from decommissioned environmentally friendly power parts. Appropriate waste administration rehearses, directed by guidelines and industry principles, are basic to limiting the ecological effect of end-of-life processes.

Inverters, being electronic gadgets, contain parts that might present ecological dangers on the off chance that not dealt with as expected during removal. Endeavors to work on the recyclability of inverters and decrease the utilization of perilous materials add to their general supportability. The execution of reclaim and reusing programs by producers guarantees that finish of-life inverters are handled in an ecologically capable way.

The roundabout economy idea, which underlines the constant use and reusing of materials to limit squander, assumes a urgent part in upgrading the manageability of sustainable power advances, including inverters.

By broadening the life expectancy of parts through restoration and reusing, the natural effect of these innovations is additionally decreased, adding to a more manageable and asset proficient energy change.

Energy capacity frameworks, frequently combined with inverters, likewise assume a part in the manageability of environmentally friendly power energy executions. Batteries utilized for energy capacity contain materials like lithium, cobalt, and nickel, which raise ecological and moral worries connected with mining practices and production network straightforwardness. Manageable obtaining of battery materials, reusing drives, and the improvement of elective sciences are areas of progressing exploration to upgrade the supportability of energy stockpiling arrangements.

The social and monetary components of supportability are basic to the general evaluation of environmentally friendly power energy frameworks. The progress to sustainable power sets out work open doors, invigorates financial development, and cultivates advancement. Networks facilitating environmentally friendly power projects, including those using inverters, frequently benefit from expanded financial action, work creation, and framework improvement.

Be that as it may, the evenhanded conveyance of the advantages of environmentally friendly power projects is a urgent thought for accomplishing social supportability. Nearby people group, especially those facilitating huge scope projects, ought to be effectively taken part in the preparation, advancement, and dynamic cycles. Local area contribution helps address concerns connected with land use, ecological effect, and guarantees that the advantages of environmentally friendly power projects are shared comprehensively.

Besides, social maintainability includes contemplations of energy access and reasonableness. Environmentally friendly power energy drives, including off-matrix arrangements fueled by inverters, add to further developing energy access in underserved locales. By giving dependable and clean power to networks without admittance to customary power frameworks, these drives upgrade social prosperity, support schooling and medical services, and enable nearby economies.

Value in the arrangement of efficient power energy advancements is critical for guaranteeing that the advantages of the energy change are available to all. Comprehensive approaches, monetary systems, and local area commitment techniques are fundamental parts of accomplishing social supportability in the sustainable power area. Addressing social imbalances and cultivating local area support add to the general achievement and acknowledgment of environmentally friendly power energy executions.

The job of strategy and administrative systems couldn't possibly be more significant in advancing natural and social manageability in the sustainable power area. Legislatures assume a focal part in forming the change to a more manageable energy scene through motivations, commands, and guidelines.

Strong arrangements, for example, feed-in levies, tax reductions, and environmentally friendly power targets, energize the reception of environmentally friendly power energy advancements, including inverters.

Administrative structures likewise assume a part in guaranteeing mindful natural practices all through the lifecycle of sustainable power frameworks. Natural effect appraisals, squander the board guidelines, and outflow principles add to limiting the negative ecological outcomes of efficient power energy projects. Straightforward and enforceable guidelines guide the feasible turn of events and activity of environmentally friendly power establishments.

Worldwide cooperation and arrangements further development the maintainability plan in the worldwide energy change. Drives, for example, the Paris Understanding highlight the significance of aggregate endeavors to address environmental change and advance manageable energy rehearses. Participation between countries, businesses, and associations facility.

5

Chapter 5

Smart Grids and Inverter Technology

Shrewd lattices and inverter innovation assume critical parts in forming the fate of energy frameworks, offering creative answers for upgrade productivity, unwavering quality, and manageability. The development of conventional power frameworks into wise and interconnected networks has become basic, given the developing interest for power, expanding joining of sustainable power sources, and the requirement for stronger foundation.

Brilliant frameworks influence progressed correspondence and data advancements to empower ongoing observing, control, and enhancement of force age, dispersion, and utilization. These frameworks work with bidirectional correspondence among utilities and end-clients, making ready for a more responsive and versatile energy foundation. With regards to brilliant matrices, inverter innovation arises as a key empowering influence, changing how energy is produced, put away, and conveyed.

One of the key difficulties looked by ordinary power lattices is their restricted ability to deal with the fluctuation and irregularity related with environmentally friendly power sources. Sun based and wind power, for example, rely upon atmospheric conditions and are dependent upon changes that can strain the steadiness of the matrix. Inverter innovation tends to this test by changing over direct flow (DC) created by sunlight powered chargers or wind turbines into substituting flow (AC), which is the standard type of power utilized in homes and organizations.

The joining of inverters considers consistent fuse of environmentally friendly power sources into the network, encouraging a more reasonable and different energy blend. As well as switching DC over completely to AC, inverters offer functionalities, for example, voltage guideline, responsive power control, and matrix synchronization. These elements upgrade the unwavering quality of sustainable power frameworks and add to matrix solidness.

Moreover, savvy lattices improve the flexibility of the general energy framework by consolidating progressed sensors, computerization, and control frameworks. These parts empower utilities to distinguish and answer shortcomings, blackouts, and different disturbances progressively. The capacity to expect and resolve issues speedily limits margin time, lessens the gamble of flowing disappointments, and works on the general unwavering quality of the power supply.

Request reaction instruments are one more critical part of brilliant frameworks. By utilizing correspondence advancements, savvy lattices empower utilities to speak with end-clients and change power utilization in light of framework conditions. This two-way correspondence engages shoppers to come to informed conclusions about their energy use, adding to by and large request decrease during top periods and upgrading the usage of assets.

In the private area, brilliant meters assume a critical part in empowering request reaction. These gadgets give continuous data on power utilization, permitting purchasers to change their use designs in view of valuing signs or framework conditions. This not just aides in lessening energy bills for purchasers yet in addition adds to the proficient usage of assets and the general dependability of the lattice.

In the domain of circulated energy assets (DERs), brilliant networks work with the consistent combination of decentralized age, energy capacity, and electric vehicles (EVs) into the framework. Inverter innovation is at the center of this incorporation, empowering the effective association of DERs to the network and giving the essential control instruments to ideal activity.

Energy capacity frameworks, combined with inverters, assume a significant part in tending to the discontinuity of sustainable power sources. By putting away abundance energy during times of high age and delivering it when request is high or age is low, energy capacity adds to matrix security and upgrades the unwavering quality of sustainable power frameworks.

With regards to electric vehicles, savvy matrices and inverter innovation support the improvement of vehicle-to-framework (V2G) frameworks. These frameworks permit electric vehicles to draw energy from the matrix as well as feed energy back into the lattice when required. This bidirectional progression of power upgrades framework adaptability, upholds top shaving, and gives an extra income stream to EV proprietors.

Network safety is a basic thought in the execution of savvy matrices. The expanded network and dependence on computerized innovations open the framework to potential digital dangers. Utilities and innovation suppliers should execute vigorous network safety measures to safeguard the respectability and secrecy of the information communicated across the framework. Encryption, verification, and interruption discovery frameworks are fundamental parts of a complete network safety technique for savvy lattices.

Interoperability and normalization are additionally key elements in the outcome of shrewd matrix organizations. Normal guidelines guarantee that various parts from different producers can flawlessly cooperate, cultivating a more open and cooperative environment. Normalization improves on combination, diminishes costs, and speeds up the reception of shrewd network innovations on a worldwide scale.

The progress to brilliant matrices includes mechanical headways as well as changes in administrative systems and plans of action. Administrative strategies need to advance to oblige the powerful idea of shrewd lattices, boost interests in innovation, and guarantee fair and impartial admittance to the advantages of framework modernization. Furthermore, utilities might have to take on new plans of action that line up with the developing job of the framework as a dynamic and intuitive stage.

While savvy matrices hold monstrous potential for working on the effectiveness and supportability of energy frameworks, their execution isn't without challenges. The forthright expenses of conveying shrewd lattice foundation can be significant, and the profit from speculation might take time. Utilities and policymakers should cautiously adjust the momentary monetary contemplations with the drawn out advantages of a more intelligent, stronger matrix.

Public mindfulness and commitment are essential parts of the fruitful sending of shrewd matrix innovations. Instructing purchasers about the advantages of shrewd networks, for example, decreased energy costs, improved unwavering quality, and ecological maintainability, can cultivate backing and collaboration. Local area association in the preparation and execution of brilliant lattice projects guarantees that the novel requirements and inclinations of various districts are considered.

Inverter innovation keeps on advancing, driven by the requirement for more noteworthy effectiveness, further developed execution, and expanded usefulness. The improvement of cutting edge inverter advances, like microinverters and power analyzers, adds to the streamlining of sun oriented power frameworks. Microinverters, for instance, are introduced on individual sunlight based chargers, permitting each board to work autonomously and augment energy creation even within the sight of concealing or glitches in different boards.

Power streamlining agents, then again, are gadgets that improve the presentation of sun oriented photovoltaic (PV) frameworks by exclusively advancing the power result of each sun based module. These advances not just further develop the general energy yield of sun oriented establishments yet additionally give more granular checking and control capacities.

The incorporation of inverter innovations with energy the board frameworks (EMS) further improves the knowledge and responsiveness of energy frameworks. EMS empowers constant checking and control of energy utilization, age, and capacity, taking into consideration dynamic changes in view of lattice conditions, energy costs, and client inclinations. This degree of control engages end-clients to effectively partake in framework improvement and add to request reaction drives.

As the sending of sustainable power frameworks keeps on growing, inverter innovations are progressively being applied to various energy sources past sun oriented and wind. For instance, inverter innovation assumes a significant part in the combination of energy stockpiling frameworks, considering effective charging and releasing of batteries. This application is especially pertinent with regards to network scale energy capacity and private energy stockpiling arrangements.

In the field of electric portability, inverters are fundamental parts of electric vehicle charging foundation. They convert AC power from the network into DC power for charging the vehicle's battery. Also, bidirectional inverters empower V2G usefulness, permitting electric vehicles to connect with the framework and add to request reaction and network adjusting.

The improvement of cutting edge inverters is additionally instrumental in upgrading the security and strength of microgrids. Microgrids are limited energy frameworks that can work freely or related to the fundamental matrix. Inverters inside microgrids empower the consistent incorporation of dispersed energy assets, like sunlight based chargers and energy stockpiling, and give the important control systems to framework security.

With regards to network modernization, the idea of the Web of Things (IoT) assumes a huge part in upgrading the network and knowledge of energy frameworks. IoT gadgets, like sensors and actuators, can be sent all through the framework foundation to assemble continuous information and empower mechanized reactions. These gadgets upgrade situational mindfulness, empower prescient upkeep, and add to the general productivity and dependability of the matrix.

AI and man-made reasoning (man-made intelligence) are progressively being utilized related to savvy frameworks to examine huge measures of information, distinguish designs, and enhance network activities. Prescient examination fueled by man-made intelligence can gauge energy interest, identify abnormalities, and upgrade the dispatch of assets for further developed effectiveness. AI calculations likewise assume a part in online protection by distinguishing and relieving possible dangers progressively.

The organization of shrewd lattices and high level inverter innovations adds to the continuous energy progress by working with the coordination of a different scope of fuel sources. Notwithstanding sun oriented and wind, shrewd networks oblige the combination of hydropower, geothermal, and other inexhaustible sources, making a stronger and practical energy environment.

As the world endeavors to lessen fossil fuel byproducts and relieve the effects of environmental change, the job of savvy frameworks in empowering a low-carbon energy future turns out to be progressively basic. The capacity of brilliant networks to oblige environmentally friendly power, advance energy proficiency, and backing the charge of transportation positions them as fundamental parts of an economical and decarbonized energy scene.

5.1 Explore the role of energy inverters in smart grid systems.

Energy inverters assume a pivotal and groundbreaking part with regards to savvy framework frameworks, filling in as a key part for the combination of environmentally friendly power sources, network soundness, and by and large energy improvement. As we dive into the investigation of their job, it becomes clear that energy inverters are at the front of modernizing power matrices, empowering a more unique, effective, and feasible energy environment.

At its center, an energy inverter is a gadget that converts direct current (DC) to substituting current (AC) or the other way around. This change is basic with regards to shrewd lattices in light of the fact that numerous sustainable power sources, like sun powered chargers and wind turbines, create power in DC. To consistently coordinate this power into the current AC-based power lattice, inverters go about as mediators, working with the smooth progression of energy between the inexhaustible sources and the network.

One of the essential difficulties looked by customary power networks is their inborn resoluteness in obliging the changeability and discontinuity related with sustainable power. Sun based and wind power age, for example, is dependent upon weather patterns and season of day, prompting vacillations in energy yield. Energy inverters address this test by adjusting the variable DC result of renewables to the predictable AC design expected for framework conveyance.

On account of sunlight based photovoltaic (PV) frameworks, inverters are instrumental in changing over the DC power produced by sun powered chargers into AC power reasonable for private or business use. This interaction is fundamental for taking care of overabundance energy back into the framework or using it locally, adding to the general proficiency and supportability of the energy framework.

Notwithstanding their basic job in changing over power, energy inverters upgrade network solidness through different high level functionalities. Voltage guideline is a key element, guaranteeing that the air conditioner power provided to the framework stays inside the predefined voltage limits. This is significant for forestalling voltage vacillations that could think twice about trustworthiness of electronic gadgets and machines associated with the lattice.

Receptive power control is one more significant part of inverter innovation. By overseeing responsive power stream, inverters add to keeping up with voltage levels and lattice soundness. This ability turns out to be especially significant within the sight of environmentally friendly power sources that may innately present changes in responsive power.

Lattice synchronization is a basic capability that guarantees that the power created by sustainable sources is in stage with the current framework. Inverters empower consistent synchronization, considering the smooth mix of environmentally friendly power without interruptions to the general framework activity.

As brilliant matrices develop, the job of energy inverters grows past basic DC to AC transformation. High level inverter innovations, like microinverters and power streamlining agents, are acquiring noticeable quality for their capacity to improve the presentation of environmentally friendly power frameworks.

Microinverters, for instance, address a shift from conventional string inverters by being introduced on individual sun powered chargers. This decentralized methodology empowers each board to work autonomously, moderating the effect of concealing or breakdowns in a single board on the general energy creation. Microinverters improve the general proficiency of sun powered PV frameworks, particularly in situations where a few boards may not get ideal daylight.

Power enhancers, then again, center around amplifying the power result of each sun oriented module. By separately enhancing the presentation of every module, power analyzers add to higher energy yields from sunlight based establishments. This granular degree of control empowers better checking, support, and generally speaking execution of the sun based power framework.

In the more extensive setting of savvy matrices, energy inverters assume a urgent part in working with the joining of conveyed energy assets (DERs). DERs envelop a different scope of decentralized energy innovations, including sunlight powered chargers, wind turbines, energy capacity frameworks, and electric vehicles (EVs). Inverter innovation turns into the normal point of interaction that empowers these assorted assets to associate consistently with the network.

Energy capacity frameworks, combined with inverters, assume a critical part in tending to the discontinuity of environmentally friendly power sources. During times of high age or low interest, overabundance energy can be put away in batteries. Conversely, during top interest or low age, put away energy can be delivered once more into the matrix. Inverter innovation guarantees the proficient charging and releasing of energy stockpiling frameworks, adding to network steadiness and unwavering quality.

The joining of electric vehicles into shrewd lattices further represents the significance of energy inverters. Inverters are fundamental parts in electric vehicle charging foundation, changing over AC power from the network into the DC power expected for charging the vehicle's battery. Additionally, bidirectional inverters empower vehicle-to-matrix (V2G) usefulness, permitting electric vehicles to consume energy from the network as well as feed energy back into the framework when required. This bidirectional stream upgrades lattice adaptability, upholds top shaving, and gives an extra income stream to electric vehicle proprietors.

Brilliant matrices, with their accentuation on constant correspondence and control, influence energy inverters to improve lattice versatility. The fuse of cutting edge sensors, computerization, and control frameworks permits utilities to recognize and answer flaws, blackouts, or variances progressively. Inverter advancements add to the dependability of environmentally friendly power frameworks by giving lattice support and empowering quick acclimations to evolving conditions.

Request reaction components, worked with by brilliant frameworks and energy inverters, address a change in perspective in how shoppers connect with and add to the network. Continuous correspondence among utilities and end-clients takes into account the change of power utilization in light of lattice conditions or valuing signals. This bidirectional correspondence enables customers to arrive at informed conclusions about their energy use, adding to request decrease during top periods and advancing asset usage.

Brilliant meters, a critical part of shrewd matrix organizations, empower request reaction at the private level. These gadgets give continuous data on power utilization, permitting buyers to screen and change their use designs. The coordination of savvy meters with energy inverters and other shrewd matrix advancements makes a powerful environment where the two utilities and buyers effectively partake in streamlining energy utilization and network execution.

Network safety arises as a basic thought in the execution of brilliant matrices, given the expanded network and dependence on computerized innovations. Energy inverters, being indispensable parts of savvy network foundation, should stick to hearty network protection measures to safeguard the trustworthiness and secrecy of information communicated across the framework. Encryption, verification, and interruption discovery frameworks are fundamental components of a thorough network protection technique for brilliant matrices.

Interoperability and normalization assume urgent parts in the progress of brilliant matrix organizations. Normal guidelines guarantee that various parts from different producers can flawlessly cooperate, cultivating an open and cooperative biological system. Normalization works on joining, lessens costs, and speeds up the reception of brilliant matrix innovations on a worldwide scale.

As energy frameworks change towards more prominent decentralization and supportability, the job of energy inverters turns out to be considerably more articulated. The capacity to proficiently change over and oversee energy from different sources, combined with cutting edge functionalities, positions energy inverters as empowering influences of a stronger and versatile energy framework.

The reception of brilliant lattices and high level inverter innovations requires a simultaneous development in administrative systems and plans of action. Administrative arrangements should adjust to oblige the unique idea of savvy matrices, boost interests in innovation, and guarantee fair and evenhanded admittance to the advantages of matrix modernization. Utilities, thusly, may have to embrace new plans of action that line up with the developing job of the matrix as a dynamic and intelligent stage.

While the likely advantages of shrewd lattices and energy inverters are critical, challenges exist in their far reaching reception. Forthright expenses related with conveying savvy framework foundation can be significant, and the profit from speculation might take time. Adjusting transient monetary contemplations with long haul benefits is a mind boggling task that utilities and policymakers should explore.

Public mindfulness and commitment are significant for the effective organization of brilliant framework innovations. Instructing buyers about the upsides of shrewd lattices, including decreased energy costs, improved unwavering quality, and ecological manageability, cultivates backing and collaboration. Including people group in the preparation and execution of brilliant matrix projects guarantees that the one of a kind requirements and inclinations of various districts are thought of, adding to a more comprehensive and responsive network.

The nonstop advancement of inverter innovation further highlights its importance in molding the eventual fate of energy frameworks. Past sun powered and wind applications, inverters are tracking down jobs in different energy areas, like energy stockpiling, electric portability, and microgrid organizations.

The improvement of more productive, minimized, and insightful inverters opens up additional opportunities for upgrading the general presentation and versatility of energy frameworks.

5.2 Discuss the benefits of smart grid technology and how inverters contribute to grid stability.

Shrewd matrix innovation addresses a huge jump forward in the development of energy frameworks, offering a scope of advantages that add to improved proficiency, dependability, and manageability. As we dig into the conversation of these advantages, it becomes clear that brilliant frameworks, combined with the critical job of inverters, hold the way to changing conventional power lattices into smart, versatile organizations.

One of the essential advantages of shrewd framework innovation lies in its capacity to improve the general productivity of energy frameworks. Through the mix of cutting edge correspondence and data advances, brilliant frameworks empower continuous checking, control, and enhancement of force age, appropriation, and utilization. This on-going perceivability into framework activities permits utilities to pursue information driven choices, prompting further developed productivity in asset portion, decreased energy misfortunes, and improved in general framework execution.

Inverters assume an essential part in accomplishing proficiency acquires inside savvy matrices, especially with regards to environmentally friendly power mix. As sustainable sources, for example, sun oriented and wind power are innately factor and irregular, inverters work with the smooth mix of these sources into the matrix. By changing over the immediate momentum (DC) yield from sunlight based chargers or wind turbines into the substituting flow (AC) utilized in conventional power lattices, inverters guarantee a consistent and normalized progression of power.

The capacity to coordinate environmentally friendly power sources is a critical driver for the reception of shrewd lattice innovation. As the interest for perfect and maintainable energy arrangements develops, savvy lattices give the system to proficiently integrating sunlight based, wind, and other renewables into the energy blend. Inverters, as the connection point between sustainable sources and the matrix, empower a more adaptable and dynamic energy framework equipped for adjusting to the changeability innate in sustainable age.

Lattice steadiness is a basic part of energy frameworks, guaranteeing that the stock of power matches the interest in a solid and predictable way. Savvy lattices, with their high level checking and control abilities, altogether add to framework security. Inverters assume a focal part in this by giving network support works, for example, voltage guideline, responsive power control, and framework synchronization.

Voltage guideline is fundamental for keeping up with the voltage levels inside endorsed limits, forestalling overvoltage or undervoltage conditions. Inverters accomplish this by progressively changing the voltage of the power they feed into the lattice. This ability turns out to be especially significant with regards to environmentally friendly power sources, which might acquaint vacillations in voltage due with their variable result.

Responsive power control is one more basic capability performed by inverters to upgrade matrix soundness. Responsive power is essential for keeping up with voltage levels and guaranteeing the proficient transmission of power. Inverters can effectively deal with the progression of responsive power, adding to generally speaking framework soundness and dependability.

Framework synchronization is worked with by inverters to guarantee that the power produced by sustainable sources is in stage with the current lattice. This synchronization is imperative for the consistent joining of sustainable power into the framework without causing disturbances. Inverters empower this synchronization interaction, considering a smooth and composed activity of the whole energy framework.

Besides, shrewd networks upgrade framework dependability through their capacity to recognize and answer deficiencies, blackouts, or changes progressively. The consolidation of cutting edge sensors, mechanization, and control frameworks permits utilities to recognize issues speedily and execute remedial measures. Inverters, as essential parts of savvy matrix foundation, add to the dependability of sustainable power frameworks by giving the important control components to fast acclimations to evolving conditions.

One more huge advantage of shrewd lattice innovation is the execution of interest reaction components. Savvy networks empower bidirectional correspondence among utilities and end-clients, considering continuous changes in power utilization in view of matrix conditions or valuing signals. This request reaction ability not just enables purchasers to come to informed conclusions about their energy use yet additionally adds to generally speaking interest decrease during top periods, enhancing the usage of assets.

In the private area, the sending of brilliant meters, related to energy inverters, works with request reaction at the singular purchaser level. Savvy meters give constant data on power utilization, empowering customers to screen and change their use designs. This communication between savvy meters and inverters makes a responsive biological system where purchasers effectively take part in enhancing their energy utilization, adding to network steadiness during times of popularity.

The mix of disseminated energy assets (DERs) is one more key advantage of savvy matrix innovation. DERs incorporate a scope of decentralized energy innovations, including sun powered chargers, wind turbines, energy capacity frameworks, and electric vehicles.

Inverters act as the point of interaction that empowers these different assets to interface consistently with the matrix, adding to a more decentralized and strong energy foundation.

Energy capacity frameworks, combined with inverters, assume a pivotal part in tending to the discontinuity of sustainable power sources. By putting away overabundance energy during times of high age and delivering it when request is high or age is low, energy capacity adds to matrix soundness and improves the dependability of environmentally friendly power frameworks. Inverters guarantee the productive charging and releasing of energy stockpiling frameworks, giving the fundamental control components to ideal activity.

Electric vehicles (EVs) address a developing part of DERs, and shrewd matrices, alongside inverters, support the improvement of vehicle-to-matrix (V2G) frameworks. V2G frameworks permit electric vehicles not exclusively to draw energy from the network yet additionally to take care of energy back into the matrix when required. This bidirectional progression of power improves framework adaptability, upholds top shaving, and gives an extra income stream to EV proprietors.

Besides, the idea of the Web of Things (IoT) assumes a huge part in improving the network and knowledge of energy frameworks inside brilliant lattices. IoT gadgets, like sensors and actuators, can be sent all through the matrix framework to assemble continuous information and empower computerized reactions. These gadgets upgrade situational mindfulness, empower prescient support, and add to the general proficiency and unwavering quality of the lattice.

AI and man-made reasoning (artificial intelligence) are progressively being utilized related to brilliant matrices to investigate huge measures of information, recognize designs, and advance network activities. Prescient examination controlled by artificial intelligence can gauge energy interest, recognize abnormalities, and upgrade the dispatch of assets for further developed proficiency. AI calculations likewise assume a part in network protection by distinguishing and relieving possible dangers continuously.

Network protection is a basic thought in the execution of savvy frameworks, given the expanded availability and dependence on computerized advances. Energy inverters, being vital parts of shrewd framework foundation, should stick to strong network safety measures to safeguard the trustworthiness and classification of information communicated across the matrix. Encryption, verification, and interruption location frameworks are fundamental components of an extensive network protection system for shrewd lattices.

Interoperability and normalization are key elements in the progress of shrewd matrix arrangements. Normal guidelines guarantee that various parts from different producers can consistently cooperate, cultivating an open and cooperative biological system. Normalization works on reconciliation, decreases costs, and speeds up the reception of brilliant network innovations on a worldwide scale.

The progress to brilliant lattices includes mechanical headways as well as changes in administrative systems and plans of action. Administrative strategies need to advance to oblige the powerful idea of savvy networks, boost interests in innovation, and guarantee fair and impartial admittance to the advantages of framework modernization. Moreover, utilities might have to take on new plans of action that line up with the developing job of the framework as a dynamic and intuitive stage.

While the advantages of shrewd matrix innovation are obvious, challenges exist in its broad reception. The forthright expenses of sending brilliant network framework can be significant, and the profit from speculation might take time. Utilities and policymakers should cautiously adjust the transient monetary contemplations with the drawn out advantages of a more astute, stronger lattice.

Public mindfulness and commitment are urgent parts of the fruitful sending of brilliant lattice advances. Instructing customers about the advantages of shrewd matrices, for example, diminished energy costs, upgraded unwavering quality, and ecological manageability, can cultivate backing and collaboration. Local area association in the preparation and execution of brilliant framework projects guarantees that the extraordinary necessities and inclinations of various districts are considered.

5.3 Address challenges and potential solutions in implementing smart grids with energy inverters.

The execution of brilliant matrices with energy inverters presents a groundbreaking way to deal with modernizing energy frameworks, however it isn't without its difficulties. As we investigate these difficulties and likely arrangements, obviously tending to specialized, administrative, and monetary obstacles is fundamental for understanding the maximum capacity of shrewd lattice innovations.

One of the essential difficulties in executing savvy matrices is the significant forthright speculation expected for sending the fundamental foundation. The establishment of cutting edge sensors, correspondence organizations, and control frameworks, alongside the coordination of energy inverters, requests critical capital consumption. This monetary hindrance can be an obstruction for utilities and legislatures, particularly when confronted with financial plan requirements or contending needs.

To defeat this test, a staged way to deal with shrewd matrix sending is frequently suggested. Utilities can focus on unambiguous locales or basic foundation regions for beginning execution, progressively growing the brilliant lattice inclusion after some time. This gradual methodology considers a more reasonable monetary weight and gives valuable chances to exhibit the advantages of savvy networks before full-scale organization.

Specialized interoperability and normalization present one more huge test in the execution of shrewd frameworks with energy inverters. The variety of innovations from different makers can prompt similarity issues and block consistent mix. Absence of normalization might bring about frameworks that battle to convey successfully, restricting the general productivity and usefulness of the savvy matrix.

One potential arrangement is to lay out and stick to normal guidelines across the business. Normalization guarantees that various parts, including energy inverters, can cooperate flawlessly. Industry associations and administrative bodies can assume an essential part in advancing and implementing these guidelines, cultivating coordinated effort among partners and smoothing out the reconciliation of different innovations.

Network protection is a basic worry in brilliant framework execution, given the expanded availability and dependence on computerized innovations. Energy inverters, being necessary parts of brilliant framework foundation, are expected focuses for digital dangers. Unapproved access, information breaks, or disturbances to matrix activities present huge dangers that should be alleviated to guarantee the security and dependability of the network.

Carrying out strong online protection measures is vital. This incorporates encryption of correspondence channels, verification conventions, and interruption location frameworks. Normal network protection reviews and updates are vital for stay in front of developing dangers. Cooperation between utilities, innovation suppliers, and network protection specialists can help create and execute viable procedures to shield savvy matrix foundation.

Administrative structures intended for customary power matrices may not be appropriate for the dynamic and decentralized nature of savvy frameworks with energy inverters. Existing guidelines might need arrangements for boosting interests in shrewd network advances, making vulnerability for utilities with respect to the recuperation of their costs. Furthermore, administrative obstacles might dial back the endorsement interaction for brilliant lattice projects.

To address these administrative difficulties, policymakers must proactively update and adjust guidelines to oblige the developing scene of brilliant matrices. This incorporates laying out clear rules for cost recuperation, making impetuses for matrix modernization, and smoothing out endorsement processes for shrewd lattice projects. Administrative systems ought to line up with the objectives of improving matrix proficiency, dependability, and maintainability.

Public obstruction and absence of mindfulness can block the execution of shrewd matrices with energy inverters. Misconceptions or concerns connected with security, wellbeing, and the potential for expanded power expenses might prompt obstruction from end-clients. Without public help, utilities might confront difficulties in conveying new advancements and drawing in shoppers sought after reaction drives.

Instructive missions and local area outreach programs are fundamental for address public worries and assemble support for shrewd matrix drives. Giving clear data about the advantages of brilliant lattices, for example, diminished energy costs, improved unwavering quality, and ecological manageability, can assist with easing fears and collect open help. Drawing in with networks in the preparation and dynamic cycles guarantees that the remarkable necessities and inclinations of various districts are thought of.

Information protection is a basic part of brilliant network execution, and worries about the assortment and utilization of individual data can frustrate public acknowledgment. Shrewd lattices create immense measures of information, including nitty gritty data about individual energy utilization designs. Safeguarding this information from unapproved access or abuse is pivotal to keeping up with public trust.

To address information security concerns, utilities and policymakers should lay out vigorous information assurance measures. This incorporates carrying out encryption conventions, severe access controls, and straightforward security approaches. Furnishing buyers with command over their information and permitting them to select in for explicit administrations can upgrade straightforwardness and fabricate trust in the shrewd lattice environment.

While energy inverters assume a pivotal part in working with the reconciliation of environmentally friendly power sources, the irregular idea of renewables presents difficulties to lattice security. Changeability in sunlight based and wind power age can prompt variances in energy supply, possibly affecting the dependability and unwavering quality of the network. This discontinuity challenge requires compelling answers for guarantee a predictable power supply.

Energy capacity frameworks, combined with energy inverters, present a feasible answer for address the discontinuity of sustainable power sources. By putting away abundance energy during times of high age and delivering it during times of low age or appeal, energy capacity frameworks add to network security. Inverters assume a focal part in dealing with the charging and releasing of energy stockpiling frameworks, guaranteeing ideal activity and upgrading in general network unwavering quality.

The mix of electric vehicles (EVs) into brilliant matrices presents extra difficulties and valuable open doors. While EVs add to the zap of transportation and can act as disseminated energy assets, the expanded interest for charging during top periods might strain the matrix. Organizing the charging of EVs to line up with network conditions and streamlining their bidirectional cooperation requires cautious administration.

Brilliant charging framework and bidirectional energy stream capacities, worked with by energy inverters, can assist with tending to the difficulties related with EV reconciliation. By utilizing V2G frameworks, electric vehicles could not just draw energy from the matrix at any point yet additionally feed overabundance energy back into the network when required.

This bidirectional stream upgrades framework adaptability, upholds top shaving, and makes extra income streams for EV proprietors.

Financial contemplations, including the expense of energy inverters and related innovations, can present difficulties to boundless reception. The underlying speculation expected for overhauling framework and sending shrewd matrix innovations, combined with the expense of cutting edge inverters, might be seen as an obstruction, especially for more modest utilities or districts with restricted monetary assets.

To address monetary difficulties, utilities can investigate supporting choices, including public-private associations, awards, and impetuses. States and administrative bodies can assume a part in offering monetary help or making motivating force projects to energize the reception of brilliant framework innovations. Exhibiting the drawn out financial advantages, including decreased functional expenses and further developed network proficiency, can assist with supporting the underlying speculation.

Discontinuous correspondence or network issues can present functional difficulties in shrewd lattice executions. The dependence on correspondence networks for continuous information trade between gadgets, sensors, and control frameworks implies that disturbances or postpones in correspondence can affect the responsiveness of the lattice. This challenge is especially pertinent in remote or topographically testing areas.

Carrying out excess correspondence organizations and reinforcement frameworks can assist with tending to availability challenges. Utilizing a mix of wired and remote correspondence innovations, alongside satellite or cell organizations, upgrades the strength of the correspondence framework. Standard support and checking of correspondence networks are fundamental to speedily distinguish and resolve issues.

The intricate idea of shrewd matrices with energy inverters requests a gifted labor force fit for planning, carrying out, and keeping up with these high level frameworks. The deficiency of experts with mastery in brilliant matrix advances, online protection, and information examination represents a human asset challenge for utilities and innovation suppliers.

Putting resources into labor force advancement programs, preparing drives, and instructive organizations can assist with tending to the abilities hole in the shrewd matrix industry. Joint efforts between scholarly organizations, industry affiliations, and service organizations can add to the improvement of a talented labor force. Constant preparation and expert advancement amazing open doors guarantee that experts keep up to date with developing innovations and best practices.

Carrying out shrewd matrices with energy inverters addresses an extraordinary undertaking that holds the possibility to reform energy frameworks, upgrade matrix effectiveness, and advance manageability. Nonetheless, this aggressive endeavor isn't without its difficulties. Tending to these difficulties and finding powerful arrangements is essential for opening the full advantages of shrewd lattice advancements and energy inverters.

One of the essential difficulties in carrying out shrewd networks with energy inverters is the critical forthright venture expected for conveying the important foundation. The establishment of cutting edge sensors, correspondence organizations, control frameworks, and the incorporation of energy inverters request significant capital consumption. This monetary obstacle can be an impediment for utilities and states, particularly when confronted with spending plan limitations or contending needs.

To moderate this test, a staged way to deal with shrewd framework sending is frequently suggested. Utilities can decisively focus on unambiguous districts or basic framework regions for beginning execution, progressively extending the savvy lattice inclusion after some time. This steady methodology takes into consideration a more reasonable monetary weight and gives potential chances to exhibit the substantial advantages of shrewd lattices prior to resolving to full-scale sending.

Specialized interoperability and normalization represent one more huge test in the execution of shrewd frameworks with energy inverters. The assorted cluster of innovations from various producers can prompt similarity issues, obstructing the consistent reconciliation of parts.

Absence of normalization might bring about frameworks that battle to impart successfully, restricting the general proficiency and usefulness of the shrewd matrix.

An expected answer for this challenge lies in the foundation and adherence to normal guidelines across the business. Normalization guarantees that various parts, including energy inverters, can cooperate consistently. Industry associations and administrative bodies assume a critical part in advancing and implementing these guidelines, cultivating joint effort among partners and smoothing out the coordination of different innovations.

Network safety arises as a basic worry in shrewd lattice execution because of the expanded availability and dependence on computerized advancements. Energy inverters, being essential parts of shrewd network foundation, are expected focuses for digital dangers. Unapproved access, information breaks, or interruptions to network tasks present huge dangers that should be addressed to guarantee the security and unwavering quality of the framework.

To handle online protection challenges, vigorous measures should be executed. This incorporates encryption of correspondence channels, severe verification conventions, and the arrangement of interruption location frameworks.

Normal online protection reviews and updates are fundamental for stay in front of developing dangers. Joint effort between utilities, innovation suppliers, and network protection specialists is essential to creating and executing powerful procedures to defend shrewd matrix framework.

Administrative systems intended for customary power matrices may not adjust flawlessly with the dynamic and decentralized nature of savvy frameworks with energy inverters. Existing guidelines might need arrangements for boosting interests in savvy framework advancements, making vulnerability for utilities in regards to the recuperation of their costs. Furthermore, administrative obstacles might dial back the endorsement cycle for savvy network projects.

To address administrative difficulties, policymakers must proactively update and adjust guidelines to oblige the advancing scene of brilliant networks. This incorporates laying out clear rules for cost recuperation, making impetuses for matrix modernization, and smoothing out endorsement processes for savvy framework projects. Administrative structures ought to line up with the objectives of improving framework effectiveness, dependability, and supportability.

Public opposition and absence of mindfulness represent extra difficulties to the execution of savvy networks with energy inverters. Errors or concerns connected with protection, wellbeing, and expected expansions in power expenses might prompt obstruction from end-clients. Without public help, utilities might experience troubles in sending new advances and drawing in buyers popular reaction drives.

Tending to public worries requires extensive instructive missions and local area outreach programs. Giving clear data about the advantages of shrewd matrices, for example, diminished energy costs, improved unwavering quality, and natural maintainability, can assist with lightening fears and earn public help. Drawing in with networks in the preparation and dynamic cycles guarantees that the novel requirements and inclinations of various districts are thought of.

Information protection is a basic part of brilliant lattice execution, and worries about the assortment and utilization of individual data can impede public acknowledgment. Shrewd networks create tremendous measures of information, including nitty gritty data about individual energy utilization designs. Shielding this information from unapproved access or abuse is significant to keeping up with public trust.

To address information security concerns, utilities and policymakers should lay out powerful information assurance measures. This incorporates carrying out encryption conventions, severe access controls, and straightforward protection strategies. Furnishing customers with command over their information and permitting them to pick in for explicit administrations can upgrade straightforwardness and construct trust in the brilliant lattice biological system.

While energy inverters assume a critical part in working with the joining of environmentally friendly power sources, the irregular idea of renewables presents difficulties to lattice solidness. Changeability in sun based and wind power age can prompt vacillations in energy supply, possibly affecting the security and unwavering quality of the lattice. This discontinuity challenge requires successful answers for guarantee a steady power supply.

Energy capacity frameworks, combined with energy inverters, present a practical answer for address the irregularity of environmentally friendly power sources. By putting away overabundance energy during times of high age and delivering it during times of low age or popularity, energy capacity frameworks add to lattice steadiness. Inverters assume a focal part in dealing with the charging and releasing of energy stockpiling frameworks, guaranteeing ideal activity and improving by and large lattice unwavering quality.

The coordination of electric vehicles (EVs) into shrewd matrices presents extra difficulties and open doors. While EVs add to the zap of transportation and can act as disseminated energy assets, the expanded interest for charging during top periods might strain the lattice. Organizing the charging of EVs to line up with lattice conditions and enhancing their bidirectional connection requires cautious administration.

Savvy charging framework and bidirectional energy stream capacities, worked with by energy inverters, can assist with tending to the difficulties related with EV combination. By utilizing Vehicle-to-Network (V2G) frameworks, electric vehicles might not just draw energy from the matrix at any point yet in addition feed overabundance energy back into the lattice when required. This bidirectional stream upgrades network adaptability, upholds top shaving, and makes extra income streams for EV proprietors.

Financial contemplations, including the expense of energy inverters and related innovations, can present difficulties to inescapable reception. The underlying speculation expected for redesigning foundation and conveying shrewd matrix innovations, combined with the expense of cutting edge inverters, might be seen as a boundary, especially for more modest utilities or districts with restricted monetary assets.

To address monetary difficulties, utilities can investigate supporting choices, including public-private associations, awards, and motivators. Legislatures and administrative bodies can assume a part in offering monetary help or making motivating force projects to energize the reception of shrewd network innovations. Exhibiting the drawn out monetary advantages, including diminished functional expenses and further developed framework productivity, can assist with supporting the underlying speculation.

Discontinuous correspondence or availability issues can present functional difficulties in shrewd network executions. The dependence on correspondence networks for constant information trade between gadgets, sensors, and control frameworks implies that disturbances or defers in

correspondence can affect the responsiveness of the matrix. This challenge is especially significant in remote or topographically testing areas.

Executing repetitive correspondence organizations and reinforcement frameworks can assist with tending to network difficulties. Utilizing a mix of wired and remote correspondence innovations, alongside satellite or cell organizations, upgrades the flexibility of the correspondence foundation. Customary support and observing of correspondence networks are fundamental to quickly recognize and resolve issues.

The complicated idea of brilliant frameworks with energy inverters requests a gifted labor force equipped for planning, executing, and keeping up with these high level frameworks. The deficiency of experts with skill in savvy framework innovations, network protection, and information examination represents a human asset challenge for utilities and innovation suppliers.

Putting resources into labor force advancement programs, preparing drives, and instructive associations can assist with tending to the abilities hole in the savvy framework industry. Coordinated efforts between scholastic foundations, industry affiliations, and service organizations can add to the improvement of a gifted labor force. Ceaseless preparation and expert advancement valuable open doors guarantee that experts keep up to date with developing innovations and best practices.

Chapter 6

Innovations in Inverter Design

Developments in inverter configuration play had a urgent impact in forming the scene of current electrical frameworks. An inverter is a gadget that converts direct flow (DC) into rotating flow (AC), considering the productive transmission and use of electrical energy. Throughout the long term, headways in innovation and a developing accentuation on manageability have driven critical developments in inverter configuration, prompting further developed execution, unwavering quality, and energy effectiveness.

One of the critical areas of advancement in inverter configuration is the joining of force hardware and semiconductor advances. Customary inverters were frequently massive and wasteful, using obsolete parts that restricted their general presentation. The approach of cutting edge semiconductor materials and assembling processes has empowered the advancement of minimal and exceptionally proficient inverters. Silicon carbide (SiC) and gallium nitride (GaN) are two instances of semiconductor materials that have reformed inverter plan.

Silicon carbide, specifically, offers predominant warm conductivity and electrical properties contrasted with customary silicon-based semiconductors. This considers higher power thickness and further developed effectiveness in inverter applications. Accordingly, inverters integrating SiC innovation can work at higher temperatures without forfeiting execution, making them ideal for requesting conditions like electric vehicles and environmentally friendly power frameworks.

Gallium nitride, then again, shows phenomenal high-recurrence qualities, empowering the plan of more modest and more lightweight inverters. GaN-based inverters are known for their quick exchanging paces and low power misfortunes, adding to by and large framework proficiency. These progressions in semiconductor innovation have prepared for the advancement of cutting edge inverters that are more smaller as well as more energy-productive.

One more huge development in inverter configuration is the utilization of cutting edge control calculations and advanced signal handling (DSP) strategies. Conventional inverters depended on straightforward control conspires that were frequently incapable to adjust to changing working circumstances. Current inverters influence complex calculations and DSP to upgrade their presentation progressively, changing boundaries like voltage, recurrence, and waveform to meet the particular prerequisites of the associated load.

Computerized signal processors assume a pivotal part in carrying out complex control calculations, considering exact tweak of the result waveform. This degree of control is fundamental in applications where the nature of the result power is basic, like in delicate electronic gadgets or clinical hardware. Additionally, high level control calculations empower inverters to flawlessly incorporate with brilliant framework innovations, working with bidirectional correspondence between the inverter and the lattice for upgraded network strength and productivity.

The reconciliation of man-made reasoning (artificial intelligence) and AI (ML) into inverter configuration addresses a notable improvement in the field. Artificial intelligence fueled inverters can constantly learn and adjust to evolving conditions, streamlining their activity in view of authentic information and continuous data sources. This versatile capacity is especially significant in sustainable power frameworks, where the result of sunlight based chargers or wind turbines can change erratically.

AI calculations can break down designs in energy creation and utilization, permitting artificial intelligence fueled inverters to expect changes and change their settings appropriately. This outcomes in better lattice combination and a more solid power supply. Also, man-made intelligence empowered inverters can upgrade shortcoming identification and diagnostics, distinguishing expected issues before they heighten and adding to the general strength of the electrical framework.

Framework shaping inverters address one more creative way to deal with inverter plan. Dissimilar to conventional framework attached inverters that require a steady matrix association with work, network shaping inverters have the capacity to lay out and keep up with lattice security all alone. This is especially pertinent in off-lattice or microgrid applications, where framework backing might be irregular or inaccessible.

Lattice framing inverters work by producing a steady AC voltage and recurrence, successfully shaping a matrix like construction. This component is significant in situations where the inverter is expected to work as the essential power source, giving energy to remote or segregated areas. The capacity of network framing inverters to establish a stable electrical climate adds to the versatility and dependability of decentralized power frameworks.

The advancement of multi-port inverters is one more significant development in the field. Customarily, inverters were intended to connect with a solitary kind of force source, like a sunlight based charger or a battery. Multi-port inverters, nonetheless, are designed to deal with various information sources at the same time, empowering more flexible and coordinated energy frameworks.

For instance, a multi-port inverter could all the while oversee inputs from sunlight powered chargers, batteries, and the primary framework, consistently exchanging between sources in light of elements like energy accessibility and cost. This adaptability is beneficial in improving energy use and decreasing dependence on a solitary power source. Multi-port inverters track down applications in a great many settings, from private sun oriented establishments to modern microgrids.

In the domain of electric vehicles (EVs), bidirectional inverters have arisen as a critical development. These inverters empower the vehicle to not just draw power from the framework for charging yet in addition to take care of overabundance energy back into the lattice when the vehicle is fixed. This bidirectional capacity is known as vehicle-to-network (V2G) innovation and is a critical part in the improvement of savvy and dynamic energy environments.

V2G innovation permits electric vehicles to go about as versatile energy stockpiling units, adding to network strength during top interest periods or crises. Bidirectional inverters in EVs empower a two-way progression of power, changing the vehicle into a circulated energy asset. This advancement not just improves the general proficiency of the electrical network yet additionally gives EV proprietors new chances to adapt their vehicle's energy stockpiling limit.

Resounding inverters address an exceptional and productive way to deal with power change. Not at all like conventional inverters that work utilizing beat width tweak (PWM), thunderous inverters influence the standards of reverberation to accomplish delicate exchanging. This decreases exchanging misfortunes and limits electromagnetic obstruction, bringing about higher effectiveness and further developed dependability.

Resounding inverters are regularly utilized in applications where high-recurrence activity is worthwhile, for example, acceptance warming, remote power move, and electric vehicle charging. The full geography considers zero-voltage exchanging, implying that the power switches can work with insignificant power dissemination. Thus, full inverters are appropriate for high-power applications where energy proficiency is a basic thought.

The reconciliation of energy stockpiling frameworks with inverters has become progressively pervasive in both private and business settings. Inverter plans that consolidate worked in energy capacity abilities empower the consistent administration of put away energy, giving reinforcement power during network blackouts and improving self-utilization of sustainable power.

These coordinated frameworks frequently include mixture inverters that can deal with both network tied and off-lattice activity modes. The energy stockpiling part, regularly as lithium-particle batteries, permits clients to store overabundance energy produced from sustainable hot-spots for sometime in the future. This adds to matrix versatility, diminishes reliance on the network, and advances more noteworthy energy autonomy.

Microinverters address a takeoff from conventional brought together inverter frameworks generally utilized in sun powered photovoltaic (PV) establishments. Rather than a solitary inverter dealing with the whole sunlight based cluster, microinverters are introduced on every individual sun powered charger. This decentralized methodology offers a few benefits, including upgraded energy collect, further developed framework unwavering quality, and improved on support.

Microinverters work freely, implying that the disappointment of one unit doesn't influence the presentation of the whole framework. This differences with brought together inverters, where a weak link can influence the whole sun powered exhibit. Furthermore, microinverters work with greatest power point following (MPPT) at the singular board level, enhancing energy collect and conquering concealing issues that might influence explicit boards.

The idea of drifting inverters has acquired consideration with regards to drifting sun oriented photovoltaic (FPV) establishments. Conventional sun oriented establishments are land-based, consuming significant earthbound space. Drifting sun powered establishments, then again, use waterways, like repositories, lakes, or even the vast ocean, to house sunlight based chargers. Drifting inverters are explicitly intended to work in the difficult ecological circumstances related with water-based sun powered establishments.

Drifting inverters should battle with variables like moistness, temperature varieties, and expected submersion. These inverters consolidate vigorous waterproofing and consumption safe highlights to guarantee dependable and sturdy activity.

The utilization of drifting sun powered establishments advances land use as well as gives extra advantages, for example, decreased water dissipation in repositories and further developed energy yield because of the cooling impact of the water.

6.1 Highlight recent advancements in inverter technology.

Late headways in inverter innovation have checked critical achievements in the development of electrical frameworks, driving enhancements in proficiency, dependability, and versatility. Inverters, gadgets that convert direct flow (DC) to exchanging flow (AC), assume a vital part in different applications, including sustainable power frameworks, electric vehicles, and matrix the executives. The most recent developments in inverter innovation range a few key regions, including semiconductor materials, control calculations, matrix communication, and joining with arising advances.

One of the eminent progressions in inverter configuration is the boundless reception of cutting edge semiconductor materials. Silicon carbide (SiC) and gallium nitride (GaN) have arisen as game-changing materials in the creation of force hardware. Silicon carbide, specifically, offers prevalent warm conductivity and electrical properties contrasted with conventional silicon-based semiconductors. This empowers higher power thickness and further developed productivity in inverter applications.

Inverters consolidating SiC innovation can work at raised temperatures without compromising execution. This capacity is especially important in requesting conditions, like those experienced in electric vehicles and environmentally friendly power frameworks. The higher productivity and power thickness of SiC-based inverters add to the general viability of these frameworks, making them more smaller and energy-effective.

Gallium nitride, another semiconductor material acquiring unmistakable quality, flaunts astounding high-recurrence attributes. Inverter plans utilizing GaN innovation benefit from quick exchanging paces and low power misfortunes. The effective exhibition of GaN-based inverters is instrumental in upgrading generally framework productivity. These progressions in semiconductor materials play had a urgent impact in the scaling down and enhancement of inverter plans, empowering them to fulfill the developing needs of current applications.

Notwithstanding progressions in semiconductor materials, the combination of cutting edge control calculations and computerized signal handling (DSP) strategies has turned into a sign of late inverter innovation. Conventional inverters depended on shortsighted control conspires that frequently attempted to adjust to changing working circumstances. Current inverters, in any case, influence complex calculations and DSP to enhance their exhibition continuously, powerfully changing boundaries like voltage, recurrence, and waveform to meet explicit burden necessities.

The execution of advanced signal processors in inverter configuration works with the exact adjustment of result waveforms, guaranteeing great power conveyance. This degree of control is critical in applications where the strength and nature of the result power are principal, like in clinical hardware or delicate electronic gadgets. Besides, high level control calculations empower consistent coordination with brilliant framework innovations, empowering bidirectional correspondence between the inverter and the matrix for further developed strength and productivity.

Man-made brainpower (man-made intelligence) and AI (ML) have likewise tracked down their direction into inverter innovation, introducing another period of versatile and smart power change. Artificial intelligence fueled inverters can persistently learn and adjust to evolving conditions, advancing their activity in view of authentic information and constant data sources. This versatile ability is especially significant in sustainable power frameworks, where the result of sunlight based chargers or wind turbines can show capricious varieties.

AI calculations examine designs in energy creation and utilization, permitting man-made intelligence fueled inverters to expect vacillations and change settings appropriately. This outcomes in superior matrix coordination and a more solid power supply. Moreover, man-made intelligence empowered inverters improve shortcoming identification and diagnostics, recognizing expected issues before they heighten, adding to the general strength of the electrical framework.

The idea of framework shaping inverters addresses a change in outlook in inverter usefulness. Dissimilar to conventional framework tied inverters that depend on a steady network association with work, lattice shaping inverters have the capacity to freely lay out and keep up with matrix steadiness. This element is especially pertinent in off-matrix or microgrid applications where lattice backing might be irregular or inaccessible.

Lattice framing inverters produce a steady AC voltage and recurrence, successfully making a matrix like construction even without any a conventional framework association. This capacity is vital in situations where the inverter is expected to work as the essential power source, giving energy to remote or detached areas. Network shaping inverters add to the flexibility and dependability of decentralized power frameworks, assuming a crucial part in the development of current microgrid models.

The improvement of multi-port inverters is another significant headway that upgrades the adaptability and joining of energy frameworks. Dissimilar to conventional inverters intended to connect with a solitary power source, multi-port inverters can deal with various sources of info at the same time. This capacity empowers more adaptable and coordinated energy frameworks, where the inverter can flawlessly switch between sources like sunlight powered chargers, batteries, and the primary matrix in view of variables like energy accessibility and cost.

In useful terms, a multi-port inverter could at the same time deal with inputs from sunlight based chargers, energy capacity frameworks, and the principal matrix, upgrading energy usage and lessening dependence on a solitary power source. This adaptability is valuable in different settings, from private sun powered establishments to modern microgrids. Multi-port inverters represent the pattern toward more versatile and dynamic power the board arrangements.

Bidirectional inverters have turned into a foundation of electric vehicle (EV) innovation, addressing a huge progression in inverter plan. These inverters empower electric vehicles not exclusively to draw power from the lattice for charging yet in addition to take care of overabundance energy back into the matrix when the vehicle is fixed. This bidirectional ability is known as vehicle-to-network (V2G) innovation and is instrumental in the advancement of shrewd and intuitive energy environments.

V2G innovation changes electric vehicles into versatile energy stockpiling units, adding to matrix solidness during top interest periods or crises. Bidirectional inverters in EVs work with a two-way progression of power, permitting the vehicle to go about as a dispersed energy asset. This development not just improves the general proficiency of the electrical framework yet in addition opens up additional opportunities for EV proprietors to use their vehicle's energy stockpiling limit with regards to monetary and ecological advantages.

Resounding inverters address an original way to deal with power change that has acquired unmistakable quality lately. Rather than customary inverters utilizing beat width tweak (PWM), full inverters exploit the standards of reverberation to accomplish delicate exchanging. This lessens exchanging misfortunes and limits electromagnetic obstruction, bringing about higher productivity and further developed dependability.

Full inverters are normally utilized in applications where high-recurrence activity is profitable, for example, enlistment warming, remote power move, and electric vehicle charging. The full geography takes into account zero-voltage exchanging, empowering power changes to work with insignificant power scattering. This makes thunderous inverters appropriate for high-power applications where energy proficiency is a basic thought.

The combination of energy stockpiling frameworks with inverters has become progressively predominant, particularly with regards to sustainable power mix. Inverter plans that consolidate worked in energy capacity abilities empower consistent administration of put away energy, giving reinforcement power during matrix blackouts and upgrading self-utilization of environmentally friendly power.

These coordinated frameworks frequently highlight half breed inverters equipped for dealing with both lattice tied and off-network activity modes. The energy stockpiling part, ordinarily made out of lithium-particle batteries, permits clients to store abundance energy produced from

inexhaustible hotspots for sometime in the future. This adds to lattice versatility, diminishes reliance on the network, and advances more prominent energy autonomy.

Microinverters address a takeoff from the conventional unified inverter frameworks pervasive in sun powered photovoltaic (PV) establishments. Rather than a solitary inverter dealing with the whole sun powered cluster, microinverters are introduced on every individual sunlight based charger. This decentralized methodology offers a few benefits, including upgraded energy collect, further developed framework unwavering quality, and worked on upkeep.

Microinverters work autonomously, implying that the disappointment of one unit doesn't affect the exhibition of the whole framework. This differentiations with brought together inverters, where a weak link can influence the whole sun based exhibit. Moreover, microinverters work with greatest power point following (MPPT) at the singular board level, advancing energy gather and moderating concealing issues that might affect explicit boards.

The idea of drifting inverters has acquired consideration, especially in the domain of drifting sunlight based photovoltaic (FPV) establishments. Customary sun powered establishments are land-based, consuming important earthbound space. Drifting sun based establishments use waterways, like repositories or lakes, to house sun powered chargers. Drifting inverters are explicitly intended to work in the difficult ecological circumstances related with water-based sun powered establishments.

6.2 Explore innovations in materials, efficiency, and miniaturization.

Lately, developments in materials, effectiveness, and scaling down have been at the front line of mechanical headways, changing different businesses and forming the fate of plan and assembling. This investigation digs into the state of the art advancements in these three basic regions, featuring how they add to improved execution, manageability, and the making of additional reduced and proficient items.

Materials Development:

High level Semiconductor Materials:

Semiconductor materials, essential in electronic gadgets and parts, have seen huge advancements. Silicon carbide (SiC) and gallium nitride (GaN) have arisen as materials that alter the productivity and execution of different electronic parts, including inverters. SiC, with its unrivaled warm conductivity and electrical properties, considers higher power thickness and further developed effectiveness in inverter applications. It succeeds in requesting conditions, like electric vehicles and sustainable power frameworks, by working productively at raised temperatures without compromising execution.

Gallium nitride, known for its magnificent high-recurrence qualities, has empowered the plan of more modest and more lightweight inverters. GaN-based inverters display quick exchanging paces and low power misfortunes, adding to in general framework productivity.

These high level semiconductor materials play had an essential impact in lessening the size, weight, and energy misfortunes in electronic parts, working with the improvement of additional minimal and superior presentation gadgets.

Lightweight and High-Strength Materials:

In enterprises going from aviation to auto, the journey for lightweight yet high-strength materials has prompted the reception of creative arrangements. High level materials like carbon fiber composites and high-strength compounds have become basic in the plan of lightweight designs that offer unrivaled strength and solidness.

Carbon fiber composites, with their outstanding solidarity to-weight proportion, have tracked down applications in aviation, where decreasing the heaviness of airplane adds to eco-friendliness. In auto fabricating, the utilization of cutting edge composites, like aluminum and titanium, has prompted the development of lighter and more eco-friendly vehicles without compromising security or primary honesty.

Shrewd Materials and Nanotechnology:

The reconciliation of brilliant materials and nanotechnology has opened new outskirts in material science. Savvy materials, which can answer outside boosts like temperature, pressure, or attractive fields, are progressively being integrated into different applications. Shape memory composites, for example, can change shape in light of temperature varieties, empowering creative plans in clinical gadgets, actuators, and aviation parts.

Nanotechnology, working at the nanoscale level, takes into account exact control of materials' properties. Nanomaterials show extraordinary mechanical, electrical, and warm attributes that vary from their full scale partners. These properties are bridled in the improvement of superior execution sensors, high level coatings, and nanocomposite materials with further developed strength and conductivity.

Productivity Progressions:

High level Control Calculations and Advanced Signal Handling:

In the domain of gadgets and power frameworks, progressions in control calculations and computerized signal handling (DSP) play had a urgent impact in further developing proficiency and execution.

Customary inverters depended on essential control plots that attempted to adjust to dynamic working circumstances. Present day inverters, in any case, influence modern calculations and DSP strategies to advance their exhibition progressively.

These high level control frameworks empower exact adjustment of result waveforms, guaranteeing excellent power conveyance. In applications where the security and nature of the result power are basic, for example, in clinical hardware or touchy gadgets, these control calculations give a critical benefit.

In addition, they work with consistent coordination with savvy framework advancements, permitting bidirectional correspondence between the inverter and the network for upgraded strength and productivity.

Computerized reasoning and AI:

The joining of man-made reasoning (simulated intelligence) and AI (ML) into electronic frameworks has introduced another period of versatile and clever usefulness. Artificial intelligence controlled frameworks can persistently learn and adjust to evolving conditions, improving their activity in view of verifiable information and constant data sources. This versatility is especially significant in sustainable power frameworks, where the result of sunlight based chargers or wind turbines can display flighty varieties.

AI calculations dissect designs in energy creation and utilization, permitting artificial intelligence fueled frameworks to expect vacillations and change settings appropriately. This further develops network joining as well as upgrades shortcoming recognition and diagnostics, adding to the general strength of electrical frameworks. Simulated intelligence and ML are progressively becoming necessary parts in electronic gadgets, upgrading productivity and versatility across different applications.

Network Framing Inverters:

The improvement of network framing inverters addresses a critical headway in inverter innovation. Conventional network attached inverters depend on a steady lattice association with work. Conversely, lattice framing inverters have the ability to autonomously lay out and keep up with framework strength. This component is especially important in off-lattice or microgrid applications where framework backing might be irregular or inaccessible.

Network framing inverters produce a steady AC voltage and recurrence, successfully making a framework like design even without any a conventional lattice association. This ability is essential in situations where the inverter is expected to work as the essential power source, giving energy to remote or disconnected areas. Lattice framing inverters add to the strength and unwavering quality of decentralized power frameworks, denoting a huge headway in network the board advancements.

Scaling down Patterns:

Multi-Port Inverters:

Ongoing improvements in inverter configuration incorporate multi-port inverters that can deal with various information sources all the while. Dissimilar to conventional inverters intended for a solitary power source, multi-port inverters consider the synchronous administration of contributions from sunlight based chargers, batteries, and the primary matrix. This adaptability streamlines energy use and lessens dependence on a solitary power source.

For example, a multi-port inverter could oversee inputs from sun powered chargers, energy capacity frameworks, and the fundamental lattice, flawlessly exchanging between sources in light of variables like energy accessibility and cost. This flexibility is useful in different settings, from private sun based establishments to modern microgrids, adding to a more coordinated and effective energy the executives framework.

Bidirectional Inverters for Electric Vehicles:

In the domain of electric vehicles (EVs), bidirectional inverters have turned into a basic part, empowering vehicles to not just draw power from the framework for charging yet additionally to take care of overabundance energy back into the network when fixed. This bidirectional capacity, known as vehicle-to-matrix (V2G) innovation, changes EVs into versatile energy stockpiling units, adding to network solidness during top interest or crises.

Bidirectional inverters in EVs work with a two-way progression of power, permitting the vehicle to go about as a conveyed energy asset. This advancement not just improves the general effectiveness of the electrical lattice yet in addition opens up additional opportunities for EV proprietors to use their vehicle's energy stockpiling limit with respect to monetary and natural advantages. The bidirectional usefulness denotes a critical stage towards making more unique and intuitive energy environments.

Resounding Inverters:

Thunderous inverters address an original way to deal with power change, especially in applications where high-recurrence activity is profitable. Not at all like conventional inverters utilizing beat width tweak (PWM), resounding inverters influence the standards of reverberation to accomplish delicate exchanging. This lessens exchanging misfortunes and limits electromagnetic obstruction, bringing about higher proficiency and further developed unwavering quality.

Thunderous inverters are ordinarily utilized in high-power applications, for example, enlistment warming, remote power move, and electric vehicle charging. The resounding geography considers zero-voltage exchanging, empowering power changes to work with insignificant power scattering. This makes thunderous inverters appropriate for situations where energy productivity is a basic thought, adding to headways in power hardware.

Incorporation of Energy Stockpiling Frameworks:

The combination of energy stockpiling frameworks with inverters has become progressively pervasive, especially with regards to environmentally friendly power coordination. Inverter plans that consolidate worked in energy capacity abilities empower the consistent administration of put away energy, giving reinforcement power during matrix blackouts and enhancing self-utilization of sustainable power.

These coordinated frameworks frequently include half and half inverters fit for dealing with both lattice tied and off-network activity modes. The energy stockpiling part, normally made out of lithium-particle batteries, permits clients to store overabundance energy created from inexhaustible hotspots for sometime in the future. This adds to network versatility, lessens reliance on the matrix, and advances more prominent energy autonomy.

Microinverters:

Microinverters address a takeoff from customary unified inverter frameworks predominant in sun based photovoltaic (PV) establishments. Rather than a solitary inverter dealing with the whole sun based cluster, microinverters are introduced on every individual sun powered charger. This decentralized methodology offers a few benefits, including upgraded energy reap, further developed framework unwavering quality, and worked on upkeep.

Microinverters work autonomously, implying that the disappointment of one unit doesn't influence the presentation of the whole framework. This differentiations with concentrated inverters, where a weak link can influence the whole sun powered exhibit. Furthermore, microinverters work with most extreme power point following (MPPT) at the singular board level, upgrading energy gather and moderating concealing issues that might affect explicit boards.

Drifting Inverters for Drifting Sun based Establishments:

The idea of drifting inverters has acquired consideration, especially with regards to drifting sun based photovoltaic (FPV) establishments. Conventional sun oriented establishments are land-based, consuming significant earthly space. Drifting sun powered establishments use waterways, like repositories or lakes, to house sunlight based chargers. Drifting inverters are explicitly intended to work in the difficult ecological circumstances related with water-based sun powered establishments.

Drifting inverters should fight with variables like stickiness, temperature varieties, and expected submersion. These inverters integrate vigorous waterproofing and erosion safe highlights to guarantee dependable and sturdy activity. The utilization of drifting sunlight based establishments advances land use and gives extra advantages, remembering diminished water vanishing for repositories and further developed energy yield because of the cooling impact of the water.

Particular Inverters:

The idea of particular inverters has gotten forward momentum in applications where adaptability and adaptability are principal. Conventional inverters are much of the time planned as solid units with fixed limits. Conversely, secluded inverters comprise of individual modules that can be joined to accomplish the ideal power rating. This particular methodology offers a few benefits, including simplicity of versatility, worked on upkeep, and further developed unwavering quality.

In a measured inverter framework, on the off chance that one module falls flat, the excess modules can keep on working, limiting personal time. This measured engineering is especially useful in huge scope applications, for example, server farms, where power prerequisites might differ, and the capacity to scale inverter limit is fundamental. Measured inverters likewise work with future updates and extensions, considering the combination of new innovations without the requirement for a total framework redesign.

Remote Power Move Advances:

The ascent of remote power move (WPT) advances has prodded development in inverter plan to help proficient and contactless energy transmission. WPT frameworks regularly utilize resounding inverter geographies to move power remotely between a source and a recipient. The inverter at the source produces a high-recurrence electromagnetic field, and the beneficiary resounds with this field to catch and change over the energy back into electrical power.

Remote inverters should meet tough effectiveness and wellbeing prerequisites, as energy move happens without actual associations. The improvement of WPT frameworks has applications in electric vehicle charging, purchaser hardware, and clinical gadgets, where the end of charging links and connectors upgrades comfort and lessens mileage.

6.3 Groundbreaking inverter designs and their impact on the industry.

Weighty inverter plans play had a groundbreaking impact in the power gadgets industry, reshaping the scene of energy transformation, dissemination, and usage. These inventive plans have presented new capacities, improved productivity, and opened up clever applications across different areas. This investigation digs into the absolute most significant weighty inverter plans and their expansive impacts on the business.

Thunderous Inverters:

One of the notable inverter plans that has gathered consideration is the resounding inverter. Not at all like conventional inverters that utilization beat width tweak (PWM), resounding inverters influence the standards of reverberation to accomplish delicate exchanging. This plan limits exchanging misfortunes and diminishes electromagnetic impedance, bringing about higher proficiency and further developed dependability.

Full inverters find applications in high-power situations, for example, enlistment warming, remote power move, and electric vehicle charging. The reverberation based approach takes into account zero-voltage exchanging, empowering power changes to work with insignificant power dissemination. The expanded productivity and decreased misfortunes add to headways in influence gadgets, particularly in applications where energy effectiveness is central.

Framework Shaping Inverters:

The improvement of lattice framing inverters denotes a critical take-off from conventional framework tied inverters.

While regular inverters depend on a steady lattice association with work, matrix shaping inverters have the capacity to freely lay out and keep up with network solidness. This is especially important in off-framework or microgrid applications where matrix backing might be irregular or inaccessible.

Network framing inverters produce a steady AC voltage and recurrence, really making a framework like design even without any a conventional lattice association. This ability is critical in situations where the inverter fills in as the essential power source, giving energy to remote or separated areas. The presentation of matrix shaping inverters adds to the flexibility and unwavering quality of decentralized power frameworks, offering a weighty answer for off-lattice power age.

Multi-Port Inverters:

Headways in inverter configuration incorporate the advancement of multi-port inverters equipped for dealing with different info sources all the while. Customary inverters were intended to communicate with a solitary power source, like a sunlight based charger or a battery. Conversely, multi-port inverters take into consideration the concurrent administration of contributions from different sources, like sunlight powered chargers, batteries, and the principal lattice.

This flexibility upgrades energy usage and diminishes dependence on a solitary power source. For instance, a multi-port inverter could oversee inputs from sunlight powered chargers, energy capacity frameworks, and the principal network, consistently exchanging between sources in light of variables like energy accessibility and cost. These multi-port inverters track down applications in assorted settings, from private sunlight based establishments to modern microgrids, exhibiting their noteworthy effect on coordinated energy frameworks.

Bidirectional Inverters for Electric Vehicles:

In the domain of electric vehicles (EVs), bidirectional inverters have arisen as a notable development. These inverters empower electric vehicles not exclusively to draw power from the framework for charging yet in addition to take care of abundance energy back into the matrix when the vehicle is fixed. This bidirectional ability, known as vehicle-to-matrix (V2G) innovation, changes EVs into versatile energy stockpiling units.

Bidirectional inverters in EVs work with a two-way progression of power, permitting the vehicle to go about as a circulated energy asset. This advancement not just improves the general effectiveness of the electrical network yet additionally opens up additional opportunities for EV proprietors to use their vehicle's energy stockpiling limit with regards to financial and ecological advantages. Bidirectional inverters address a critical stage towards making more unique and intuitive energy biological systems, with suggestions for both the auto and energy businesses.

Joining of Energy Stockpiling Frameworks:

The joining of energy stockpiling frameworks with inverters has turned into an earth shattering pattern, particularly with regards to environmentally friendly power reconciliation. Inverter plans that consolidate worked in energy capacity abilities empower consistent administration of put away energy, giving reinforcement power during network blackouts and upgrading self-utilization of environmentally friendly power.

These coordinated frameworks frequently include half breed inverters fit for dealing with both network tied and off-lattice activity modes. The energy stockpiling part, commonly made out of lithium-particle batteries, permits clients to store overabundance energy produced from sustainable hotspots for sometime in the future. This noteworthy combination adds to network flexibility, decreases reliance on the framework, and advances more prominent energy freedom.

Microinverters:

Microinverters address a takeoff from customary concentrated inverter frameworks normally utilized in sunlight based photovoltaic (PV) establishments. Rather than a solitary inverter dealing with the whole sunlight based exhibit, microinverters are introduced on every individual sun powered charger. This decentralized methodology offers a few earth shattering benefits, including upgraded energy reap, further developed framework unwavering quality, and improved on upkeep.

Microinverters work autonomously, implying that the disappointment of one unit doesn't affect the presentation of the whole framework. This differentiations with brought together inverters, where a weak link can influence the whole sun oriented exhibit. Also, microinverters work with greatest power point following (MPPT) at the singular board level, streamlining energy gather and relieving concealing issues that might influence explicit boards. The reception of microinverters addresses a change in outlook in sun oriented PV establishments, opening more noteworthy effectiveness and unwavering quality.

Drifting Inverters for Drifting Sunlight based Establishments:

The idea of drifting inverters has acquired noticeable quality, especially with regards to drifting sun powered photovoltaic (FPV) establishments. Customary sun oriented establishments are land-based, consuming significant earthly space. Drifting sun based establishments use waterways, like repositories or lakes, to house sun powered chargers. Drifting inverters are explicitly intended to work in the difficult ecological circumstances related with water-based sun oriented establishments.

Drifting inverters should battle with variables like dampness, temperature varieties, and possible submersion. These inverters integrate hearty waterproofing and erosion safe highlights to guarantee dependable and strong activity.

The utilization of drifting sun based establishments streamlines land use as well as gives extra advantages, remembering diminished water vanishing for repositories and further developed energy yield because of the cooling impact of the water. Drifting inverters address a notable answer for growing sunlight based establishments to water bodies, introducing new open doors for feasible energy age.

Particular Inverters:

The idea of particular inverters has built up momentum in applications where versatility and adaptability are vital. Conventional inverters are many times planned as solid units with fixed limits. Conversely, particular inverters comprise of individual modules that can be consolidated to accomplish the ideal power rating. This particular methodology offers noteworthy benefits, including simplicity of versatility, worked on upkeep, and further developed unwavering quality.

In a measured inverter framework, in the event that one module falls flat, the leftover modules can keep on working, limiting personal time. This secluded engineering is especially gainful in enormous scope applications, for example, server farms, where power prerequisites might change, and the capacity to scale inverter limit is fundamental. Measured inverters likewise work with future redesigns and extensions, taking into consideration the coordination of new innovations without the requirement for a total framework upgrade. The reception of measured inverters addresses an earth shattering movement towards more adaptable and versatile power foundation.

Remote Power Move Innovations:

The ascent of remote power move (WPT) advances has prodded historic developments in inverter plan to help effective and contactless energy transmission. WPT frameworks commonly utilize full inverter geographies to move power remotely between a source and a collector. The inverter at the source creates a high-recurrence electromagnetic field, and the recipient resounds with this field to catch and change over the energy back into electrical power.

Remote inverters should meet rigid effectiveness and security prerequisites, as energy move happens without actual associations. The improvement of WPT frameworks has pivotal applications in electric vehicle charging, purchaser hardware, and clinical gadgets, where the disposal of charging links and connectors upgrades comfort and lessens mileage. The reception of remote power move innovations addresses a noteworthy stage towards a link free future, reforming the manner in which we send and get electrical energy.

Chapter 7

Energy Storage and Inverters

Energy capacity and inverters assume critical parts in present day energy frameworks, filling in as fundamental parts that empower effective and solid energy the board. As the world keeps on changing towards sustainable power sources, the requirement for viable energy stockpiling arrangements turns out to be progressively critical. This requires a more profound comprehension of the standards, innovations, and applications related with energy capacity frameworks and inverters.

Energy capacity fills in as a key part for the joining of environmentally friendly power sources, for example, sunlight based and wind, into the power matrix. These sources are innately discontinuous, with energy creation subject to factors like atmospheric conditions and season of day. Subsequently, energy capacity gives a way to catch overabundance energy during times of overflow and delivery it when request is high or when inexhaustible sources are not effectively creating power.

One of the essential difficulties in the energy stockpiling space is the determination of appropriate stockpiling advancements. Different innovations exist, each with its extraordinary attributes, benefits, and impediments. Batteries are among the most predominant energy stockpiling arrangements, incorporating different sciences like lithium-particle, lead-corrosive, and stream batteries.

Lithium-particle batteries, specifically, have acquired conspicuousness in applications going from electric vehicles to private energy stockpiling frameworks. Their high energy thickness, long cycle life, and generally low self-release make them appropriate for catching and delivering energy effectively. Nonetheless, challenges like asset accessibility, wellbeing concerns, and natural effect highlight the significance of continuous examination to upgrade lithium-particle battery innovation.

Lead-corrosive batteries, with their long history of purpose, stay important for specific applications. They are practical and dependable, making them appropriate for uninterruptible power supply (UPS) frameworks and reinforcement power applications. Notwithstanding, their lower energy thickness and more limited cycle life contrasted with lithium-particle batteries limit their far reaching use in arising energy stockpiling markets.

Stream batteries address one more class of energy stockpiling innovations that address specific impediments of customary batteries. Stream batteries use substance intensifies disintegrated in fluid electrolytes, empowering versatile and adaptable energy stockpiling arrangements. Vanadium redox stream batteries, for example, have been investigated for network scale applications because of their capacity to decouple power and energy limit, giving more prominent functional adaptability.

Past batteries, other energy stockpiling innovations incorporate packed air energy capacity (CAES), siphoned hydro capacity, and nuclear power stockpiling. CAES frameworks store energy by compacting air and putting away it in underground caves or compartments. During times of appeal, the compacted air is delivered to drive turbines and create power. Siphoned hydro capacity includes siphoning water to a raised repository during seasons of surplus energy and delivering it to a lower supply through turbines when power is required.

Nuclear power stockpiling frameworks store and delivery energy as intensity. This can be accomplished through stage change materials, reasonable intensity stockpiling, or inert intensity stockpiling. These frameworks find applications in sun based nuclear energy stations, where intensity can be put away during radiant periods and used to create power when daylight is lacking.

Inverters assume a critical part in energy frameworks by changing over direct current (DC) into substituting current (AC) as well as the other way around. In environmentally friendly power frameworks, inverters are fundamental for coordinating power from sources like sunlight based chargers into the air conditioner matrix. Sunlight based photovoltaic (PV) boards produce DC power, and inverters convert this DC power into AC power viable with the network.

The two fundamental kinds of inverters are focal inverters and string inverters. Focal inverters are bigger units that handle the transformation of DC to AC for a whole sunlight based exhibit. They are many times utilized in utility-scale sunlight based establishments. Then again, string inverters are more modest units associated with individual strings of sun powered chargers. They are usually utilized in private and business sun based establishments. Each type enjoys its benefits and contemplations, like productivity, cost, and upkeep necessities.

Microinverters address a later development in sun oriented power frameworks. Not at all like focal and string inverters, microinverters are introduced straightforwardly on individual sun powered chargers. This decentralized methodology offers benefits like expanded framework productivity, worked on support, and the capacity to autonomously screen the presentation of each sunlight based charger. Nonetheless, the expense and intricacy of microinverter frameworks might influence their inescapable reception.

Notwithstanding sun oriented inverters, there are inverters intended for other environmentally friendly power sources, like breeze turbines. Wind inverters convert the variable and frequently offbeat AC power produced by wind turbines into framework viable AC power. Similarly as with sun based inverters, wind inverters assume a basic part in guaranteeing the smooth mix of wind energy into the power network.

The reconciliation of energy stockpiling with inverters improves the general exhibition and adaptability of energy frameworks. Batteries, for instance, can be combined with inverters to store overabundance energy and release it when required. This capacity is essential for tending to the irregular idea of sustainable power sources and guaranteeing a steady and dependable power supply.

High level inverter innovations, like brilliant inverters, add to the knowledge and versatility of present day energy frameworks. Savvy inverters have the ability to speak with the matrix and different gadgets, empowering highlights like voltage guideline, recurrence control, and lattice support. These functionalities upgrade the dependability of the power lattice and work with the reconciliation of a higher level of sustainable power.

The idea of a "savvy matrix" remains closely connected with the development of energy stockpiling and inverter innovations. A savvy matrix use progressed correspondence and control innovations to streamline the age, dispersion, and utilization of power. Energy capacity, with its capacity to store and delivery energy on a case by case basis, turns into a key empowering influence for a more brilliant and more proficient matrix.

In a shrewd lattice situation, energy capacity frameworks outfitted with smart inverters can add to stack evening out, top shaving, and request reaction. Load evening out includes streamlining changes in energy interest, guaranteeing a more steady and unsurprising burden on the matrix. Top shaving alludes to the decrease of power utilization during times of appeal, assisting with staying away from network overburdens and the requirement for extra power age limit.

Request reaction includes changing power utilization in light of signs from the framework administrator. Energy capacity frameworks, constrained by brilliant inverters, can assume a urgent part in answering these signs by either putting away overabundance energy during times of low interest or releasing put away energy during top interest periods.

This advantages the network administrator as well as permits customers to exploit lower power costs during off-top hours.

The charge of transportation acquaints one more aspect with the job of energy stockpiling and inverters. Electric vehicles (EVs) depend on energy capacity frameworks, commonly lithium-particle batteries, to store and convey electrical energy for impetus. In this unique circumstance, inverters are fundamental for changing over DC power from the vehicle's battery into AC ability to drive the electric engine.

The mix of EVs into the framework raises open doors and difficulties. From one perspective, EVs can act as versatile energy stockpiling units, adding to network strength through vehicle-to-framework (V2G) innovation. V2G permits EVs to release overabundance energy back to the framework when required, successfully transforming them into circulated energy assets. Then again, the expanded interest for power coming about because of broad EV reception requires cautious preparation and the board of the matrix to stay away from over-burdens.

Lattice scale energy capacity projects are turning out to be more predominant as the requirement for huge scope environmentally friendly power coordination develops. These ventures frequently include the arrangement of gigantic battery frameworks, like the Tesla Megapack, to store and delivery gigawatt-long stretches of energy. The blend of cutting edge energy capacity advancements and inverters with framework scale abilities adds to the strength and unwavering quality of the power network.

As energy stockpiling and inverter innovations keep on propelling, innovative work endeavors are centered around tending to key difficulties and pushing the limits of execution. For energy capacity, one basic area of exploration includes the advancement of new materials and sciences to upgrade the energy thickness, cycle life, and wellbeing of batteries.

Developments in stream battery advances, for instance, mean to defeat current constraints and give versatile answers for matrix scale energy capacity.

The journey for feasible and harmless to the ecosystem energy capacity arrangements likewise incorporates investigating elective advances. Strong state batteries, for example, address a promising road for what's to come. These batteries supplant conventional fluid electrolytes with strong materials, offering potential benefits like higher energy thickness, further developed security, and longer cycle life. Be that as it may, the commercialization of strong state batteries at scale requires defeating specialized and producing difficulties.

In the domain of inverters, research endeavors center around further developing proficiency, unwavering quality, and framework similarity. The improvement of cutting edge power hardware, including wide-bandgap semiconductors like silicon carbide (SiC) and gallium.

7.1 Discuss the synergy between energy storage systems and inverters.

The collaboration between energy capacity frameworks and inverters is at the front of changing the scene of energy the executives and power dissemination. This cooperative relationship is fundamental to tending to the difficulties presented by the coordination of environmentally friendly power sources, streamlining lattice dependability, and encouraging a stronger and practical energy foundation.

Energy capacity frameworks, enveloping various advances like batteries, siphoned hydro capacity, and warm stockpiling, act as essential parts in the mission for a cleaner and more dependable energy future. These frameworks assume a basic part in catching and putting away overabundance energy created during times of high environmentally friendly power creation, considering its delivery during times of expanded request or when sustainable sources are not effectively producing power.

The joining of inverters into this situation is similarly indispensable. Inverters, which convert direct momentum (DC) into rotating ebb and flow (AC) as well as the other way around, empower the consistent incorporation of sustainable power into the power network. With regards to sustainable power sources like sunlight based chargers and wind turbines, inverters assume a pivotal part in changing over the variable and frequently irregular DC power produced by these sources into the stable and lattice viable AC power.

The collaboration between energy capacity frameworks and inverters turns out to be especially obvious in situations where the discontinuous idea of environmentally friendly power sources presents difficulties to keeping a steady power supply. For example, sunlight powered chargers create power just when presented to daylight, and wind turbines produce power when the breeze is blowing.

This discontinuity presents vacillations in energy creation, making it important to store abundance energy during top creation periods and delivery it when request is high or when sustainable sources are not effectively producing power.

Batteries, among the most common energy stockpiling arrangements, are frequently combined with inverters to accomplish this goal. During times of high environmentally friendly power creation, overabundance power is coordinated to charge the batteries. In this charging system, the DC power created by sun powered chargers or wind turbines is changed over into AC power by the inverters prior to being put away in the batteries. At the point when request surpasses the ongoing environmentally friendly power creation or when sustainable sources are not effectively producing power, the put away energy in the batteries is delivered, and the inverters convert it back to AC power for appropriation to the network or end-clients.

This unique exchange between energy capacity frameworks and inverters adds to lattice dependability and unwavering quality, relieving the difficulties related with the irregular idea of sustainable power sources. By putting away overabundance energy and delivering it when required, this collaboration guarantees a more reliable and unsurprising power supply, tending to one of the key worries related with the boundless reception of environmentally friendly power.

Additionally, the blend of energy stockpiling frameworks and inverters improves the general adaptability of energy the board. Energy capacity frameworks outfitted with smart inverters can answer matrix flags and adjust their activity in light of constant interest and supply conditions. This ability adds to stack evening out, where variances in energy request are streamlined, and top shaving, where power utilization is decisively decreased during top interest periods.

The idea of interest reaction is one more region where the collaboration between energy capacity and inverters assumes a vital part. Request reaction includes changing power utilization because of signs from the matrix administrator. Energy capacity frameworks, constrained by savvy inverters, can effectively take part popular reaction programs by regulating their activity in light of matrix signals. This not just advantages the general steadiness of the power framework yet additionally permits shoppers to participate in more financially savvy energy utilization rehearses.

The canny coordination worked with by this collaboration reaches out past simple energy stockpiling and dissemination. Savvy inverters, equipped for correspondence with the lattice and different gadgets, add to the general knowledge of the power framework. They can effectively participate in voltage guideline, recurrence control, and framework support, subsequently upgrading the steadiness and versatility of the power network.

The idea of a "savvy framework" rises out of these progressions, wherein computerized advancements, correspondence organizations, and high level energy stockpiling frameworks with canny inverters work couple to make a more effective and responsive power foundation.

In a savvy lattice situation, the mix of energy stockpiling and inverters considers the streamlining of energy age, circulation, and utilization, introducing a time of energy the board described by expanded effectiveness and supportability.

With regards to environmentally friendly power, the collaboration between energy capacity and inverters is essential for conquering a portion of the innate difficulties related with these energy sources. Sun oriented and wind power, for example, are dependent upon varieties in atmospheric conditions and season of day. The capacity to store overabundance energy during times of high creation and delivery it during times of low creation helps address the irregularity of these sustainable sources.

Siphoned hydro capacity is one more energy stockpiling innovation that embodies the collaboration with inverters. In this strategy, overabundance power is utilized to siphon water from a lower repository to an upper supply during seasons of high environmentally friendly power creation. At the point when power request is high or when sustainable sources are not effectively producing power, the put away water is set free from the upper repository to the lower supply, going through turbines to create power. In this cycle, inverters assume an essential part in changing over the mechanical energy from the turbines into electrical energy reasonable for the framework.

Nuclear power stockpiling frameworks, depending on the standards of putting away and delivering energy as intensity, likewise benefit from the collaboration with inverters. In sun oriented nuclear energy stations, for instance, overabundance sun powered energy can be utilized to warm a stockpiling medium during radiant periods. At the point when daylight is deficient, the put away intensity can be delivered to create steam and drive turbines, creating power. Inverters are fundamental in changing the nuclear power over completely to power, going full circle of energy transformation and use.

The development of energy stockpiling and inverter innovations additionally converges with the charge of transportation. Electric vehicles (EVs), outfitted with energy capacity frameworks as batteries, depend on inverters to change over DC power from the battery into AC power for driving the electric engine. This combination of energy stockpiling and inverters in the domain of transportation further features the flexibility and cross-cutting uses of these advances.

The idea of vehicle-to-lattice (V2G) innovation adds one more layer to the collaboration between energy capacity and inverters with regards to electric vehicles. V2G empowers electric vehicles to draw power from the matrix as well as to take care of overabundance energy back into the network when left and associated. This bidirectional progression of power requires complex inverter innovations that can flawlessly switch among charging and releasing modes, adding to lattice solidness and supporting interest reaction drives.

At the matrix scale, enormous energy stockpiling projects are turning out to be progressively pervasive. These tasks frequently include the arrangement of huge battery frameworks, like the Tesla Megapack, to store and delivery gigawatt-long periods of energy. The mix of cutting edge inverters with network scale capacities guarantees the smooth activity and coordination of these energy stockpiling projects with the more extensive power lattice. This collaboration adds to the security and dependability of the lattice while obliging the developing portion of environmentally friendly power sources.

Chasing after propelling energy stockpiling and inverter innovations, progressing innovative work endeavors are fundamental. One basic area of examination includes upgrading the presentation and effectiveness of batteries. This incorporates investigating new materials and sciences to further develop energy thickness, cycle life, and security. Developments in stream battery advancements, strong state batteries, and elective sciences add to the nonstop development of energy stockpiling arrangements.

In the space of inverters, research centers around further developing productivity, unwavering quality, and lattice similarity. High level power hardware, including wide-bandgap semiconductors like silicon carbide (SiC) and gallium nitride (GaN), assume a urgent part in improving the effectiveness of inverters. These materials empower higher exchanging frequencies, diminish power misfortunes, and add to the general exhibition of energy transformation.

The idea of "virtual power plants" (VPPs) is picking up speed as a way to total and streamline the presentation of circulated energy assets, including energy stockpiling frameworks and inverters. VPPs influence computerized advancements and correspondence organizations to arrange the activity of different energy resources, expanding their aggregate effect on network solidness and unwavering quality. This approach lines up with the more extensive pattern towards decentralized and versatile energy frameworks.

7.2 Explore how inverters contribute to efficient energy storage and retrieval.

Inverters assume a significant part in the domain of energy stockpiling and recovery, filling in as fundamental parts that work with the consistent change and usage of put away energy. This investigation digs into the multi-layered manners by which inverters add to the effectiveness of energy stockpiling frameworks, going from their association in the mix of sustainable power sources to their job in overseeing and advancing the recovery of put away energy.

At the center of this conversation is the connection point between direct current (DC) and exchanging current (AC). Inverters, by definition, are electronic gadgets intended to change over DC power into AC power or the other way around. This ability is especially huge with regards to energy capacity frameworks, where the put away energy is normally as DC power, and the matrix or end-clients frequently require AC power for utilization.

One of the essential spaces where inverters feature their importance is in the combination of sustainable power sources, like sunlight based chargers and wind turbines, into the power framework. These sources innately produce DC power, and it is the job of inverters to change over this DC power into the air conditioner power that lines up with the standard network foundation. Sun based photovoltaic (PV) boards, for instance, create power as DC power when presented to daylight. Inverters are utilized to change over this DC power into AC power, empowering the consistent infusion of sun oriented energy into the lattice.

The effectiveness of this change cycle is a basic figure the general presentation of environmentally friendly power frameworks. Inverters should be intended to limit power misfortunes during the DC to AC transformation, guaranteeing that the greatest measure of energy reaped from sunlight based chargers or wind turbines is really taken care of into the framework. Propels in inverter advances, for example, the utilization of wide-bandgap semiconductors like silicon carbide (SiC) and gallium nitride (GaN), add to higher productivity by diminishing exchanging misfortunes and improving the general influence change process.

With regards to energy capacity, inverters accept a similarly essential job. Batteries, one of the most widely recognized types of energy stockpiling, store electrical energy as DC power. At the point when the put away energy should be used, inverters convert the DC power from the batteries into AC power, making it viable with the framework or reasonable for direct use by electrical machines and gadgets. This change cycle is critical for guaranteeing the consistent coordination of energy stockpiling frameworks into the more extensive energy foundation.

Productive energy stockpiling and recovery include something other than the transformation of DC to AC power. Inverters are instrumental in dealing with the charge and release patterns of energy stockpiling frameworks, advancing the general execution and life span of batteries. Battery the executives frameworks (BMS) coordinated with inverters assume a pivotal part in observing variables like condition of charge, voltage, and temperature, guaranteeing that the batteries work inside protected and proficient boundaries.

In addition, inverters add to the mental fortitude of energy stockpiling frameworks. Shrewd inverters, outfitted with correspondence abilities, empower continuous observing and control of energy stockpiling units. This network considers dynamic changes because of matrix conditions, energy interest, and other outer elements. The capacity to speak with the framework and different gadgets positions brilliant inverters as vital participants popular reaction drives, where energy utilization can be progressively changed in view of matrix signals.

The idea of "matrix framing inverters" addresses a change in outlook in the job of inverters in energy capacity. Customarily, inverters have been viewed as network following gadgets, changing their activity to match the voltage and recurrence of the matrix.

Interestingly, framework shaping inverters can independently lay out and keep up with the voltage and recurrence of the network. This capacity upgrades the security and versatility of the power framework, particularly in situations where matrix associations might be feeble or missing.

In off-framework or far off regions, where energy capacity frameworks frequently work freely, network shaping inverters add to the foundation of microgrids. Microgrids are restricted energy frameworks that can work freely or related to the primary matrix. Framework shaping inverters assume a vital part in the development and synchronization of microgrids, guaranteeing that the energy produced and put away inside these frameworks is really used to satisfy nearby need.

The productivity of energy stockpiling and recovery isn't exclusively subject to the specialized determinations of inverters; it likewise depends on the decision of energy stockpiling advancements. Different energy stockpiling frameworks show assorted attributes as far as energy thickness, release rates, cycle life, and productivity. Inverters should be customized to the particular necessities of every energy stockpiling innovation to expand generally speaking framework productivity.

Lithium-particle batteries, generally utilized in different energy stockpiling applications, require inverters that can deal with the particular voltage and current attributes of these batteries. In addition, headways in inverter advancements for lithium-particle batteries include developments in power gadgets, control calculations, and warm administration to improve by and large framework productivity and security.

Stream batteries, one more class of energy stockpiling innovation, present exceptional difficulties and valuable open doors for inverters. These batteries store energy in fluid electrolytes and, in contrast to conventional batteries, decouple power and energy limit. Inverters for stream batteries should be intended to deal with the variable and frequently lower voltage levels related with these frameworks. Moreover, the capacity to proficiently deal with the progression of fluid electrolytes is a critical thought in the plan of inverters for stream battery applications.

The combination of cutting edge materials, for example, strong state electrolytes, in energy capacity frameworks presents new contemplations for inverters. Strong state batteries offer likely benefits with regards to somewhere safe, energy thickness, and cycle life. Inverters for strong state batteries should be adjusted to the particular qualities of these advances, guaranteeing ideal execution and dependability.

Past the domain of conventional batteries, other energy stockpiling advancements, like compacted air energy capacity (CAES) and nuclear power stockpiling, additionally depend on inverters for productive energy change.

CAES frameworks, for example, include the pressure and development of air to store and delivery energy. Inverters assume a basic part in changing over the mechanical energy from air venture into electrical energy for framework infusion.

On account of nuclear power stockpiling, where energy is put away and delivered as intensity, inverters are utilized to change over the nuclear power into power. This is exemplified in concentrated sun oriented power (CSP) plants, where mirrors center daylight to warm a liquid, and the subsequent nuclear power is utilized to produce steam and drive turbines. Inverters in these frameworks are answerable for changing over the mechanical energy from turbines into electrical energy viable with the network.

As energy stockpiling frameworks keep on developing, the interest for inverters that can flawlessly coordinate with an assortment of capacity innovations turns out to be progressively articulated. The flexibility and versatility of inverters assume a urgent part in supporting the organization of different energy stockpiling arrangements, taking care of the particular necessities of various applications and matrix designs.

The productivity of energy stockpiling and recovery additionally crosses with the more extensive idea of energy framework flexibility. Versatility alludes to the capacity of an energy framework to endure and recuperate from disturbances, like cataclysmic events or matrix disappointments. In this unique situation, the job of inverters stretches out past proficient energy change; they become key components in guaranteeing the solidness and unwavering quality of energy frameworks during antagonistic circumstances.

Microgrid applications, where energy capacity and neighborhood age are joined to make independent energy frameworks, feature the strength upgrading abilities of inverters. In case of a matrix blackout, inverters in microgrids can empower consistent changes to islanded activity, where the microgrid works freely, providing capacity to basic burdens. This capacity improves the general versatility of the energy framework by decreasing reliance on a brought together lattice.

The effectiveness of energy stockpiling and recovery is likewise impacted by the financial matters of the general energy framework. The expense adequacy of inverters, related to energy capacity advances, decides the practicality of far reaching organization. Propels in inverter advances that add to higher productivity, diminished support costs, and further developed unwavering quality improve the monetary practicality of energy stockpiling projects.

The idea of full circle productivity is a key measurement in assessing the effectiveness of energy stockpiling frameworks. It measures the proportion of the energy yield during release to the energy input during charging, taking into account misfortunes brought about during the capacity and recovery process.

Inverters altogether add to full circle proficiency, and continuous innovative work endeavors center around working on this measurement to make energy capacity frameworks all the more financially serious.

The combination of computerized reasoning (artificial intelligence) and AI calculations with inverters further upgrades the effectiveness of energy stockpiling and recovery. These advances empower prescient investigation and continuous improvement, permitting inverters to adjust their activity in light of verifiable information, weather conditions figures, and framework conditions.

7.3 Highlight emerging technologies and trends in energy storage coupled with inverters.

The scene of energy stockpiling combined with inverters is consistently developing, driven by innovative headways, market elements, and the basic to make a more supportable and versatile energy future. This investigation dives into arising advancements and patterns that are forming the direction of energy stockpiling and inverters, from creative capacity answers for insightful inverters and their mix into brilliant matrices.

One striking pattern in the domain of energy stockpiling is the developing conspicuousness of cutting edge battery advances. While lithium-particle batteries have overwhelmed the market, specialists and producers are investigating elective sciences and materials to address limits, for example, asset shortage, wellbeing concerns, and natural effect. Strong state batteries, specifically, stand out enough to be noticed as a possible major advantage. These batteries supplant the customary fluid electrolyte with strong materials, offering benefits concerning security, energy thickness, and cycle life. Inverters intended for strong state batteries need to adjust to the novel attributes of these innovations, and progressing research means to streamline the incorporation of strong state batteries with productive and wise inverters.

Stream batteries address one more arising innovation in the energy stockpiling scene. Not at all like customary batteries, stream batteries store energy in fluid electrolytes, considering versatile and adaptable arrangements. Vanadium redox stream batteries have built up some decent momentum for framework scale applications because of their capacity to decouple power and energy limit, giving improved functional adaptability. Inverters intended for stream batteries assume a pivotal part in dealing with the variable voltage levels related with these frameworks and improving the transformation of electrical energy during charge and release cycles.

Additionally, the investigation of elective sciences, for example, sodium-particle batteries and zinc-air batteries, presents additional opportunities for energy capacity. Sodium-particle batteries, with their wealth of unrefined components, are being investigated as a more feasible option in contrast to lithium-particle batteries.

Zinc-air batteries, utilizing the response among zinc and oxygen, offer the potential for high energy thickness. Inverters viable with these arising battery advances should be adequately flexible to deal with various voltage ranges and release qualities, driving development in power hardware and control frameworks.

Past progressions in battery innovations, the mix of computerized reasoning (man-made intelligence) and AI into energy capacity and inverter frameworks is a groundbreaking pattern. Man-made intelligence driven energy the executives frameworks improve the productivity and unwavering quality of energy stockpiling by anticipating energy interest, enhancing charge and release cycles, and adjusting to dynamic network conditions. Savvy inverters outfitted with simulated intelligence capacities add to lattice security through continuous checking, prescient investigation, and versatile control methodologies. The collaboration among simulated intelligence and energy stockpiling combined with inverters is preparing for more insightful and versatile energy frameworks.

With regards to inverters, lattice framing inverters address a change in outlook. Generally, inverters have been network following gadgets, changing their activity to match the voltage and recurrence of the matrix. Lattice shaping inverters, then again, have the ability to independently lay out and keep up with matrix voltage and recurrence. This upgrades the steadiness and flexibility of energy frameworks, particularly in situations where matrix associations are feeble or missing. The joining of lattice framing inverters lines up with the more extensive pattern of making more independent and independent microgrids, especially in remote or off-matrix areas.

The idea of virtual power plants (VPPs) is acquiring conspicuousness as a way to total and streamline the exhibition of dispersed energy assets, including energy stockpiling frameworks and inverters. VPPs influence computerized innovations and correspondence organizations to arrange the activity of assorted energy resources, boosting their aggregate effect on matrix security and unwavering quality. This approach lines up with the more extensive pattern towards decentralized and strong energy frameworks. Inverters inside VPPs assume a crucial part in guaranteeing the planned and proficient activity of dispersed energy assets, adding to the general adaptability and unwavering quality of the power matrix.

Besides, the charge of transportation is impacting the joining of energy stockpiling frameworks and inverters. Electric vehicles (EVs), outfitted with energy capacity as batteries, present open doors for bidirectional energy stream. Vehicle-to-network (V2G) innovation permits EVs to draw power from the lattice as well as feed abundance energy back into the framework when left and associated. Inverters that flawlessly switch among charging and releasing modes assume an essential part in supporting V2G applications. This bidirectional progression of energy improves the general productivity and adaptability of the matrix while offering extra benefit to EV proprietors through potential income age.

The pattern towards particular and versatile energy stockpiling arrangements is building up some decent momentum, driven by the requirement for adaptability and flexibility in assorted applications. Measured energy capacity frameworks, made out of individual units that can be effortlessly associated or disengaged, offer benefits with regards to adaptability, simplicity of support, and cost-adequacy. Inverters intended for measured energy capacity frameworks should oblige changing arrangements and flawlessly incorporate with various sorts of capacity units. This pattern lines up with the developing idea of energy frameworks, where the capacity to increase or down in view of interest is a pivotal thought.

In the journey for supportability, the investigation of second-life applications for batteries is arising as an essential pattern. As batteries from electric vehicles arrive at the finish of their valuable life for vehicular applications, they actually hold a critical piece of their ability. Reusing these batteries for fixed energy capacity applications expands their life expectancy and diminishes the natural effect of battery removal. Inverters adjusted for second-life battery applications need to represent the attributes of matured batteries and enhance their presentation inside the limitations of diminished limit and possible debasement.

Energy capacity combined with inverters is progressively turning into a vital piece of catastrophe strong foundation. The capacity to store energy privately, combined with keen inverters, guarantees a solid power supply during framework blackouts brought about by cataclysmic events or different crises. Microgrids furnished with energy capacity frameworks and lattice shaping inverters can work independently, giving basic capacity to fundamental administrations, for example, clinics, crisis havens, and correspondence organizations. This pattern lines up with the more extensive objective of improving the versatility of energy foundation notwithstanding unexpected disturbances.

With regards to sustainable power coordination, the idea of half and half environmentally friendly power frameworks is getting some momentum. These frameworks join various environmentally friendly power sources, for example, sun oriented and wind, with energy capacity and keen inverters. Cross breed frameworks offer benefits as far as expanded dependability, upgraded energy catch, and further developed lattice combination. Inverters inside cross breed frameworks assume a pivotal part in dealing with the variable results of various sustainable sources, streamlining energy transformation, and guaranteeing a steady and synchronized association with the lattice.

The investigation of energy stockpiling innovations with expanded release span is another pattern forming the business. While conventional batteries succeed in applications requiring short explosions of energy, there is a developing requirement for capacity arrangements fit for giving supported control overstretched periods. Inverters for such applications should be intended to deal with the novel release attributes of these stockpiling innovations, taking into account productive and constant energy conveyance. This pattern is especially pertinent in situations where dependable power supply is fundamental for off-lattice or far off applications.

The combination of energy stockpiling, inverters, and blockchain innovation is an arising pattern with likely ramifications for decentralized energy frameworks. Blockchain innovation works with secure and straightforward distributed exchanges, empowering new models of energy exchanging and sharing. Inverters inside such frameworks assume a part in dealing with the bidirectional progression of energy between members, guaranteeing the exactness and security of exchanges. This pattern lines up with the vision of making more fair and decentralized energy organizations.

The unique scene of energy stockpiling combined with inverters is described by a large number of patterns that are reshaping the manner in which we outfit, store, and disseminate energy. These patterns, driven by mechanical developments, market requests, and the worldwide push towards supportable energy arrangements, altogether add to the continuous advancement of the energy stockpiling and inverter industry.

One unmistakable pattern in the field of energy stockpiling is the fast headway and reception of lithium-particle batteries. These batteries have turned into the foundation of numerous energy stockpiling applications because of their high energy thickness, somewhat lengthy cycle life, and far and wide business accessibility. Inverters intended for lithium-particle batteries assume a urgent part in proficiently changing over the immediate flow (DC) put away in these batteries into substituting flow (AC), making it viable with the power framework.

Notwithstanding, the dependence on lithium-particle innovation has likewise started a mission for choices. The push for manageability, combined with worries about the restricted accessibility of lithium assets, has prompted the investigation of new battery sciences. Strong state batteries, for instance, address a promising other option. By supplanting the fluid electrolyte with a strong material, these batteries offer expected benefits concerning wellbeing, energy thickness, and life expectancy. Inverters customized for strong state batteries need to oblige the exceptional qualities of these frameworks, provoking continuous innovative work in the field of force hardware.

One more arising pattern in energy capacity is the ascent of stream batteries. These frameworks store energy in fluid electrolytes contained in outside tanks, considering decoupling of force and energy limit. Vanadium redox stream batteries, specifically, definitely stand out for network scale applications because of their adaptability and capacity to give expanded release term. Inverters in stream battery applications should fight with the variable voltage levels related with these frameworks and upgrade the change of electrical energy during charge and release cycles.

Past batteries, elective energy stockpiling advancements are likewise gaining ground. Packed air energy capacity (CAES), for example, includes putting away energy by compacting air and delivering it when expected to produce power. Inverters assume a vital part in changing over the mechanical energy from air venture into electrical energy reasonable for lattice infusion.

Additionally, gravitational energy stockpiling, where potential energy is put away by lifting weighty masses and delivering them to produce power, is acquiring consideration. Inverters in such frameworks should effectively change over the mechanical energy from the falling masses into electrical energy.

The incorporation of man-made consciousness (computer based intelligence) and AI (ML) into energy capacity and inverter frameworks addresses an extraordinary pattern. Simulated intelligence driven energy the board frameworks improve the proficiency and dependability of energy stockpiling by anticipating energy interest, streamlining charge and release cycles, and adjusting to dynamic network conditions. Brilliant inverters furnished with simulated intelligence capacities add to network soundness through constant checking, prescient investigation, and versatile control methodologies. The collaboration among artificial intelligence and energy stockpiling combined with inverters is making ready for more insightful and versatile energy frameworks.

Lattice framing inverters are another outlook changing pattern in the inverter space. Generally, inverters have worked as network following gadgets, changing their result to match the voltage and recurrence of the framework. Lattice shaping inverters, nonetheless, have the capacity to independently lay out and keep up with matrix voltage and recurrence. This upgrades the strength and versatility of energy frameworks, particularly in situations where network associations are feeble or missing. The joining of lattice shaping inverters lines up with the more extensive pattern of making more independent and independent microgrids, especially in remote or off-network areas.

Virtual power plants (VPPs) are acquiring unmistakable quality as a way to total and upgrade the presentation of dispersed energy assets, including energy stockpiling frameworks and inverters. Utilizing advanced innovations and correspondence organizations, VPPs coordinate the activity of assorted energy resources, augmenting their aggregate effect on network solidness and dependability. Inverters inside VPPs assume a crucial part in guaranteeing the organized and effective activity of disseminated energy assets, adding to the general adaptability and dependability of the power network.

The jolt of transportation is affecting the reconciliation of energy stockpiling frameworks and inverters. Electric vehicles (EVs), outfitted with energy capacity as batteries, present open doors for bidirectional energy stream. Vehicle-to-matrix (V2G) innovation permits EVs to draw power from the framework as well as feed abundance energy back into the lattice when left and associated. Inverters that consistently switch among charging and releasing modes assume a critical part in supporting V2G applications. This bidirectional progression of energy improves the general proficiency and adaptability of the framework while offering extra benefit to EV proprietors through potential income age.

The pattern towards measured and versatile energy stockpiling arrangements is building up some forward movement. Particular energy stockpiling frameworks, made out of individual units that can be effortlessly associated or detached, offer benefits regarding versatility, simplicity of support, and cost-viability. Inverters intended for measured energy capacity frameworks should oblige changing setups and consistently incorporate with various kinds of capacity units. This pattern lines up with the developing idea of energy frameworks, where the capacity to increase or down in view of interest is a pivotal thought.

Supportability stays a main impetus in the energy stockpiling and inverter scene. The investigation of second-life applications for batteries is an arising pattern with critical ecological ramifications. As batteries from electric vehicles arrive at the finish of their valuable life for vehicular applications, they actually hold a critical piece of their ability. Reusing these batteries for fixed energy capacity applications broadens their life expectancy and diminishes the natural effect of battery removal. Inverters adjusted for second-life battery applications need to represent the attributes of matured batteries and improve their exhibition inside the limitations of decreased limit and expected corruption.

Energy capacity combined with inverters is progressively turning into a fundamental piece of catastrophe strong foundation. The capacity to store energy privately, combined with savvy inverters, guarantees a dependable power supply during lattice blackouts brought about by catastrophic events or different crises. Microgrids furnished with energy capacity frameworks and lattice shaping inverters can work independently, giving basic capacity to fundamental administrations, for example, medical clinics, crisis havens, and correspondence organizations. This pattern lines up with the more extensive objective of improving the versatility of energy foundation despite unanticipated interruptions.

Cross breed environmentally friendly power frameworks are acquiring noticeable quality as a pattern that joins various environmentally friendly power sources with energy capacity and insightful inverters. These frameworks influence the reciprocal idea of various sustainable sources, for example, sunlight based and wind, to upgrade in general unwavering quality and lattice joining. Inverters inside mixture frameworks assume a significant part in dealing with the variable results of various sustainable sources, streamlining energy change, and guaranteeing a steady and synchronized association with the network.

The investigation of energy stockpiling innovations with broadened release length is another pattern molding the business. While conventional batteries succeed in applications requiring short eruptions of energy, there is a developing requirement for capacity arrangements fit for giving supported control overstretched periods. Inverters for such applications should be intended to deal with the interesting release attributes of these stockpiling advances, taking into consideration effective and ceaseless energy conveyance. This pattern is especially pertinent in situations where dependable power supply is fundamental for off-network or far off applications.

The combination of energy stockpiling, inverters, and blockchain innovation is an arising pattern with possible ramifications for decentralized energy frameworks. Blockchain innovation works with secure and straightforward distributed exchanges, empowering new models of energy exchanging and sharing. Inverters inside such frameworks assume a part in dealing with the bidirectional progression of energy between members, guaranteeing the exactness and security of exchanges. This pattern lines up with the vision of making more fair and decentralized energy organizations.

Chapter 8

Challenges and Future Trends

The world is continually developing, driven by innovative headways, cultural changes, and worldwide difficulties. In this unique scene, different areas face remarkable difficulties, and exploring through them requires imaginative arrangements and a proactive methodology. This paper investigates the difficulties and future patterns across various spaces, going from innovation and medical services to natural supportability and then some.

Innovation and Development:

The quick speed of mechanical development is both an aid and a test. On one hand, forward leaps in man-made consciousness (computer based intelligence), AI, and mechanical technology vow to alter businesses, upgrade productivity, and work on our personal satisfaction. Then again, the speed of innovative headway frequently outperforms our capacity to control and adjust to these changes.

Network safety stands apart as a basic test in the mechanical scene. As we become progressively reliant upon interconnected frameworks and advanced stages, the danger of cyberattacks poses a potential threat.

The assurance of touchy data, protected innovation, and basic framework requests steady watchfulness and versatile safety efforts.

Besides, the moral ramifications of arising advancements bring up complex issues. Issues like information security, algorithmic inclination, and the effect of robotization on business need cautious thought. Finding some kind of harmony between innovative advancement and moral obligation is a test that requires cooperative endeavors from legislatures, enterprises, and the general population.

Medical services and General Wellbeing:

The medical services area faces multi-layered difficulties, with a developing worldwide populace, the ascent of persistent sicknesses, and the continuous danger of irresistible illnesses. The Coronavirus pandemic featured the weaknesses in our worldwide medical care foundation, stressing the requirement for readiness, flexibility, and global cooperation.

Admittance to medical care stays a critical test, with differences in medical care conveyance and results enduring across districts and financial gatherings. Spanning these holes requires inventive arrangements, including telemedicine, portable wellbeing applications, and local area based medical care drives.

Besides, the rising pervasiveness of antimicrobial obstruction represents a serious danger to general wellbeing. The abuse of anti-toxins and other antimicrobial specialists adds to the advancement of safe types of microscopic organisms, making once-treatable contaminations more challenging to make due. Addressing this challenge requires a planned worldwide work to advance capable antimicrobial use and put resources into the improvement of new helpful procedures.

Ecological Supportability:

The planet faces an existential emergency as the outcomes of environmental change become progressively clear. Climbing worldwide temperatures, outrageous climate occasions, and the deficiency of biodiversity highlight the dire requirement for thorough ecological manageability measures.

The progress to environmentally friendly power sources is a critical part of tending to environmental change. The reliance on petroleum products contributes essentially to ozone harming substance outflows, driving a worldwide temperature alteration. Putting resources into sun powered, wind, and other environmentally friendly power innovations, alongside further developing energy effectiveness, is significant for relieving the effect of environmental change.

Preservation of regular assets and the insurance of biological systems are similarly basic parts of natural maintainability. Deforestation, overfishing, and contamination endanger the sensitive equilibrium of environments, influencing biodiversity and biological system administrations. Carrying out supportable practices in agribusiness, ranger service, and fisheries is fundamental to safeguarding the strength of the planet.

Social and Financial Imbalance:

The persevering test of social and monetary disparity stays a considerable hindrance to worldwide advancement. Differences in pay, schooling, and admittance to potential open doors make a pattern of drawback for underestimated networks. Tending to these disparities requires a multi-layered approach that incorporates monetary strategies, training changes, and civil rights drives.

Orientation disparity is an inescapable issue that keeps on influencing social orders around the world. Shutting the orientation pay hole, advancing equivalent portrayal in positions of authority, and testing imbued generalizations are fundamental stages toward accomplishing orientation equality. Moreover, guaranteeing admittance to training and medical services for ladies and young ladies is urgent for breaking the pattern of orientation based separation.

Comprehensive monetary development is essential for lessening by and large disparity. Strategies that focus on work creation, support little and medium-sized ventures, and cultivate business add to building stronger and evenhanded economies. Social wellbeing nets and government assistance projects can give a support against monetary shocks and assist with lifting weak populaces out of destitution.

Segment Movements and Urbanization:

The worldwide populace is going through critical segment shifts, with suggestions for economies, medical care frameworks, and social designs. Maturing populaces in many created nations present difficulties connected with medical care and annuity frameworks, while in a few creating districts, fast populace development presents difficulties as far as offering essential types of assistance and foundation.

Urbanization is a characterizing pattern of the 21st hundred years, with additional individuals moving to urban areas looking for better open doors. While urbanization can drive financial development and advancement, it likewise strains metropolitan framework, prompting issues like gridlock, deficient lodging, and natural corruption. Reasonable metropolitan preparation and the improvement of savvy urban communities are significant for guaranteeing that urbanization adds to generally prosperity.

Worldwide Administration and Collaboration:

Tending to the difficulties of the 21st century requires successful worldwide administration and global collaboration.

Issues, for example, pandemics, environmental change, and digital dangers rise above public boundaries, requiring cooperative endeavors to track down arrangements.

Nonetheless, the ongoing international scene presents impediments to worldwide participation. Strains between significant powers, exchange debates, and an ascent in patriot feelings challenge the idea of a brought together worldwide local area. Conquering these difficulties requires discretionary drives, compromise components, and a promise to multilateralism.

The job of worldwide associations, like the Assembled Countries, becomes vital in cultivating collaboration and tending to worldwide difficulties. Reinforcing these foundations and guaranteeing their viability in planning reactions to emergencies are basic for building a stronger and interconnected world.

Instruction and Long lasting Learning:

The quick speed of mechanical change and the developing idea of work request a change in our way to deal with training. Customary models of schooling may not satisfactorily get ready people for the abilities required in the 21st-century labor force. Underlining STEM (science, innovation, designing, and math) training, computerized proficiency, and decisive reasoning abilities is fundamental for furnishing people with the devices they need to prevail in a quickly impacting world.

Besides, the idea of deep rooted learning is acquiring unmistakable quality as a reaction to the developing idea of work. Ceaseless upskilling and reskilling become significant for people to stay cutthroat in the gig market. Bosses, instructive foundations, and policymakers need to team up to make adaptable learning pathways that oblige the different necessities of students all through their lives.

Moral Contemplations in Innovation:

As innovation keeps on progressing, moral contemplations become progressively significant. The turn of events and arrangement of man-made intelligence, biotechnology, and other state of the art innovations bring up significant moral issues. Guaranteeing that these innovations are utilized to serve mankind and don't fuel existing imbalances requires insightful moral structures and guidelines.

One area of concern is the moral utilization of artificial intelligence, especially in dynamic cycles. The potential for predisposition in calculations, absence of straightforwardness, and unseen side-effects present dangers to people and society at large. Laying out rules for moral man-made intelligence advancement and organization is fundamental to forestall hurt and guarantee mindful development.

Moreover, the moral ramifications of hereditary designing and biotechnology request cautious investigation. The capacity to control the human genome brings up issues about the limits of clinical intercession, the potential for fashioner children, and the drawn out ramifications for human advancement. Moral discussions encompassing these innovations need to include an expansive range of partners, including researchers, ethicists, policymakers, and the overall population.

The Eventual fate of Work:

The idea of work is going through a significant change, driven via mechanization, computer based intelligence, and other mechanical headways. While these progressions hold the commitment of expanded effectiveness and new open doors, they likewise raise worries about work relocation, pay imbalance, and the requirement for a re-imagined common agreement.

Computerization can possibly dispense with specific routine errands, prompting the making of new positions that require remarkably human abilities, like inventiveness, decisive reasoning, and the capacity to understand people on a profound level. Notwithstanding, the change might be trying for people in enterprises vigorously affected via mechanization. Strategies that help labor force retraining, reskilling projects, and social security nets are significant for guaranteeing a smooth change.

8.1 Address current challenges in the field of energy inverters.

The field of energy inverters is a basic part of present day energy frameworks, assuming a urgent part in changing over and overseeing electrical power. As the interest for spotless and supportable energy sources keeps on rising, energy inverters face a bunch of difficulties that should be addressed to guarantee the effectiveness, unwavering quality, and life span of sustainable power frameworks. This conversation will dig into a portion of the ongoing difficulties in the field of energy inverters, investigating mechanical, financial, and natural perspectives that influence their exhibition.

One of the essential difficulties in the domain of energy inverters is the requirement for further developed effectiveness. Inverters are fundamental parts in sustainable power frameworks, changing over direct current (DC) produced by sunlight based chargers or wind turbines into substituting current (AC) for use in homes and organizations. The proficiency of this change cycle is essential for amplifying the general energy result of these frameworks. Customary inverters frequently face effectiveness misfortunes because of variables like exchanging misfortunes, conduction misfortunes, and warm misfortunes.

Exchanging misfortunes happen during the progress between the on and off conditions of semiconductor gadgets inside the inverter. As the interest for higher exchanging frequencies increments to upgrade power thickness and decrease the size of inverters, tending to exchanging misfortunes turns out to be considerably more basic.

Analysts and specialists are effectively dealing with creating progressed semiconductor materials and plans to limit exchanging misfortunes and further develop by and large inverter productivity.

Conduction misfortunes, originating from the opposition of the inverter parts, additionally add to decreased proficiency. Advancements in materials and assembling processes are being investigated to make parts with lower obstruction, subsequently limiting conduction misfortunes. Furthermore, the utilization of wide-bandgap semiconductors, like silicon carbide (SiC) and gallium nitride (GaN), shows guarantee in lessening conduction misfortunes and working on the general execution of energy inverters.

Warm misfortunes represent another critical test. Inverters produce heat during activity, and unnecessary temperatures can prompt diminished proficiency and long haul unwavering quality issues. Powerful warm administration frameworks are fundamental to disperse heat and keep up with ideal working circumstances. Incorporating progressed cooling advancements, for example, fluid cooling or imaginative intensity sinks, can assist with alleviating warm misfortunes and upgrade the general exhibition and life expectancy of energy inverters.

One more basic test in the field of energy inverters is the issue of matrix mix. As sustainable power sources like sun oriented and wind become progressively common, the inconstancy and irregularity of their result present difficulties for the steadiness and dependability of electrical lattices. Energy inverters assume a key part in framework joining by dealing with the fluctuating power yield from sustainable sources and guaranteeing a smooth and stable stockpile of power to the network.

Lattice codes and guidelines change across districts, and energy inverters should consent to these norms to work with consistent joining into existing power framework. This requires progressed control calculations and correspondence frameworks inside inverters to answer progressively to network conditions. Savvy lattice advancements, including continuous observing and correspondence abilities, are being integrated into inverters to improve network soundness and unwavering quality.

Moreover, the developing pattern of appropriated energy assets (DERs), like roof sunlight based chargers and private energy stockpiling frameworks, adds intricacy to network joining. Inverters should be furnished with the knowledge to oversee bidirectional power streams, permitting energy to be both taken care of into the matrix and drawn from it in light of interest and network conditions. This bidirectional capacity requires modern control procedures and correspondence conventions to guarantee the consistent mix of DERs into the more extensive energy scene.

Monetary contemplations likewise assume a critical part in the difficulties looked by energy inverters. The underlying expense of inverters and related parts stays an obstruction to far reaching reception of environmentally friendly power frameworks.

While the expense of sunlight powered chargers and wind turbines has seen an extensive lessening lately, inverter costs have not diminished at a similar rate. This cost dynamic can restrict the monetary reasonability of environmentally friendly power projects and dial back the progress to a more practical energy foundation.

To address this test, research endeavors are centered around creating practical inverter advancements without compromising execution. The utilization of novel materials, smoothed out assembling cycles, and economies of scale underway are a portion of the methodologies being investigated to decrease the general expense of energy inverters. Furthermore, progressions in power hardware and semiconductor advancements can possibly drive down expenses and make sustainable power more serious with customary energy sources.

Dependability and toughness are central worries in the field of energy inverters. These gadgets work in assorted and in some cases unforgiving natural circumstances, going from outrageous temperatures to high mugginess and destructive environments. The life span of inverters is vital for the monetary suitability of sustainable power projects, as incessant support and substitutions can bring about huge expenses and upset energy creation.

Working on the unwavering quality of energy inverters includes upgrading the toughness of parts and streamlining framework plans. Powerful nooks and defensive coatings can protect inverters from natural elements, while thorough testing and quality control measures can recognize and address potential disappointment focuses. In addition, the reception of prescient upkeep methods, empowered by sensors and checking frameworks, can help expect and forestall issues before they lead to framework disappointments.

Network safety represents a developing worry in the energy area, and energy inverters are not resistant to the dangers related with computerized dangers. As inverters become more associated and coordinated into shrewd matrix frameworks, they become possible focuses for cyberattacks that could think twice about trustworthiness and security of force organizations. Guaranteeing the online protection of energy inverters is critical for keeping up with the trust and versatility of the whole energy foundation.

Creating strong online protection measures requires cooperation between industry partners, government offices, and network safety specialists. Encryption conventions, secure correspondence channels, and normal programming refreshes are fundamental parts of a far reaching network safety procedure for energy inverters. Furthermore, cultivating a culture of network safety mindfulness and giving preparation to experts in the energy area can assist with moderating the dangers related with digital dangers.

The joining of energy stockpiling frameworks with inverters presents the two potential open doors and difficulties. Energy capacity innovations, for example, batteries, assume a crucial part in relieving the discontinuity of environmentally friendly power sources and working on the

general unwavering quality of energy frameworks. Nonetheless, the combination of capacity presents intricacies as far as framework configuration, control methodologies, and monetary contemplations.

One test is improving the coordination among inverters and energy stockpiling to boost the proficiency of charging and releasing cycles. This requires complex control calculations that consider factors, for example, energy interest, framework conditions, and the condition of charge of the stockpiling framework. Furthermore, the financial feasibility of energy stockpiling arrangements relies upon elements, for example, battery expenses, life expectancy, and administrative systems.

Administrative structures and principles assume a urgent part in molding the scene for energy inverters. Changing norms and affirmation prerequisites across areas can make difficulties for producers and ruin the interoperability of inverters in various business sectors. Fitting worldwide principles and smoothing out confirmation cycles can work with the worldwide sending of energy inverters and advance a more predictable and productive sustainable power framework.

8.2 Discuss potential solutions and advancements on the horizon.

In tending to the difficulties looked by energy inverters, a few likely arrangements and headways are not too far off, offering promising roads for development in effectiveness, lattice joining, monetary suitability, unwavering quality, and network safety.

Effectiveness stays a point of convergence for development in energy inverters. Headways in semiconductor materials and configuration are expected to assume a critical part in working on the proficiency of energy transformation processes. The utilization of wide-bandgap semiconductors, like silicon carbide (SiC) and gallium nitride (GaN), presents a potential chance to limit exchanging misfortunes and lessen generally speaking energy misfortunes in inverters. These materials offer prevalent warm properties and empower higher working temperatures, adding to expanded proficiency and dependability.

In addition, research endeavors are centered around creating progressed cooling advancements to address warm misfortunes. Fluid cooling frameworks, improved heat sinks, and inventive warm administration arrangements expect to disperse heat all the more successfully, guaranteeing ideal working circumstances for energy inverters. By relieving warm misfortunes, these progressions upgrade effectiveness as well as add to expanding the life expectancy of inverters, lessening upkeep costs, and further developing generally framework unwavering quality.

Network mix arrangements are imperative for adjusting energy inverters to the advancing scene of force frameworks. Shrewd matrix innovations, combined with cutting edge control calculations, work with consistent joining of sustainable power sources into the lattice.

Continuous checking, correspondence capacities, and matrix responsive inverters empower dynamic acclimations to changing lattice conditions, improving security and dependability.

The fuse of bidirectional power stream capacities is fundamental for overseeing conveyed energy assets (DERs). Inverters furnished with bidirectional usefulness can proficiently take care of overabundance energy into the lattice during times of surplus age and draw power from the network when inexhaustible sources are deficient. This bidirectional capacity requires refined control techniques to enhance energy streams and keep up with framework soundness, particularly in situations with a high entrance of DERs.

To address financial contemplations, progressing research expects to diminish the expense of energy inverters through different methodologies. Smoothed out assembling processes, economies of scale, and the utilization of savvy materials add to bringing down creation costs. Proceeded with headways in power gadgets and semiconductor advances are expected to prompt more reasonable parts, making sustainable power frameworks monetarily serious with customary sources.

Dependability and solidness arrangements include working on the power of energy inverters to endure different natural circumstances. Advancements in defensive coatings, solid walled in areas, and materials with high protection from erosion add to the life span of inverters. Prescient support procedures, empowered by sensors and checking frameworks, give early identification of possible issues, taking into account opportune intercessions and limiting the gamble of framework disappointments.

The mix of energy stockpiling frameworks with inverters opens additional opportunities for tending to irregularity challenges. High level control calculations that streamline the coordination among inverters and energy stockpiling improve the proficiency of charging and releasing cycles. Besides, progressing headways in energy capacity advances, like upgrades in battery sciences and decreases in costs, add to the monetary suitability of coordinated arrangements.

Online protection measures are fundamental to shielding energy inverters from computerized dangers. Encryption conventions, secure correspondence channels, and ordinary programming refreshes are indispensable parts of an extensive network protection technique. Joint effort between industry partners, government organizations, and online protection specialists is crucial for stay in front of advancing dangers and guarantee the versatility of energy framework.

Looking forward, arising advances hold the possibility to upset the field of energy inverters. Quantum figuring, for example, could improve the computational capacities of inverters, empowering more refined control calculations and advancement techniques. Man-made brainpower (computer based intelligence) and AI (ML) calculations are being investigated to anticipate and adjust to dynamic network conditions, working on the general execution of energy frameworks.

The idea of "self-mending" inverters is additionally acquiring consideration. Inverters with self-analytic abilities can recognize and redress issues independently, diminishing the dependence on manual upkeep and further developing framework unwavering quality. Self-mending innovations influence ongoing information, simulated intelligence calculations, and high level sensors to quickly identify irregularities and execute remedial activities.

In the domain of materials science, progressing research investigates novel materials with improved electrical and warm properties. Materials that can work at higher temperatures without compromising execution could add to more productive and dependable energy inverters. Moreover, headways in nanotechnology might prompt the advancement of parts with further developed conductivity and decreased opposition, further improving inverter execution.

In the mission for manageable energy arrangements, bi-directional energy move advances are getting some momentum. These innovations permit energy to stream from inexhaustible sources to the matrix as well as from the framework to energy capacity or different burdens. Bi-directional energy move upgrades the adaptability of energy frameworks, taking into account better variation to changing interest and network conditions.

The idea of "energy switches" addresses a potential change in perspective in energy the board. Energy switches would wisely coordinate power streams inside a decentralized energy framework, improving the utilization of sustainable power sources, energy capacity, and the matrix. These gadgets would assume a vital part in adjusting organic market, boosting the use of sustainable power, and guaranteeing a versatile and productive energy foundation.

Moreover, developments in nanogrids and microgrids offer confined answers for energy challenges. Nanogrids, comprising of limited scope appropriated energy assets, can work autonomously or associate with the principal matrix. Microgrids, which include a more extensive scope of fuel sources and loads, give confined flexibility and can detach from the primary lattice during crises. The two ideas add to upgraded energy security and manageability.

As the progress to a more practical energy future speeds up, strategy and administrative structures will assume a significant part in molding the scene for energy inverters. Smoothing out worldwide principles and confirmation cycles can work with the worldwide sending of energy inverters, encouraging interoperability and consistency across various business sectors. Clear and steady strategies that boost the reception of sustainable power advancements add to a better climate for development and venture.

8.3 Explore future trends and the role of inverters in the evolving energy landscape.

The fate of the energy scene is set apart by a unique exchange of innovative progressions, strategy changes, and a developing accentuation on manageability. In this developing setting, energy inverters are set to assume a urgent part in forming the direction of the energy area. Investigating future patterns and the developing job of inverters gives experiences into the groundbreaking changes that will characterize the following period of the worldwide energy progress.

One of the conspicuous future patterns in the energy scene is the proceeded with development of environmentally friendly power sources. Sun powered and wind energy, specifically, are encountering quick extension as their costs decline and effectiveness moves along. This flood in sustainable power sending requires progressed energy inverters to change over and deal with the variable result from these sources proficiently.

Energy capacity advances are ready to become essential parts of the energy scene, tending to the irregularity related with renewables. Inverters will assume a focal part in the mix of energy stockpiling frameworks, empowering consistent charging and releasing cycles. The collaboration among inverters and energy stockpiling improves network adaptability, upholds lattice solidness, and empowers a more dependable and tough energy framework.

As the charge of transportation picks up speed, the interest for electric vehicles (EVs) is supposed to take off. This shift towards electric portability presents new difficulties and open doors for energy frameworks. Inverters will be critical in working with the bidirectional progression of force among EVs and the lattice, empowering effective charging and releasing cycles. Savvy inverters outfitted with cutting edge correspondence capacities will assume an imperative part in improving the charging of EVs in view of matrix conditions and power interest.

The idea of shrewd lattices is an extraordinary pattern that imagines a canny and interconnected energy framework. In this unique circumstance, inverters act as shrewd hubs inside the lattice, able to do ongoing correspondence and dynamic changes. Brilliant inverters assume a key part in improving lattice security, overseeing dispersed energy assets, and empowering two-way correspondence between energy makers and customers.

The ascent of decentralized energy frameworks, including nanogrids and microgrids, addresses a change in perspective in energy age and conveyance. Inverters become integral to the activity of these decentralized frameworks by dealing with the different blend of fuel sources, like sunlight based chargers, wind turbines, and energy stockpiling, at a confined level. This pattern enables networks and organizations to have more noteworthy command over their energy supply, cultivating energy versatility and manageability.

Digitalization and the incorporation of cutting edge innovations, like computerized reasoning (man-made intelligence) and the Web of Things (IoT), are ready to upset the energy area. Brilliant inverters outfitted with man-made intelligence calculations can streamline energy creation and utilization, foresee matrix conditions, and adjust to changing conditions progressively. The sending of IoT gadgets related to inverters takes into consideration improved observing, information assortment, and controller, adding to more proficient and responsive energy frameworks.

With regards to future energy drifts, the job of energy inverters reaches out past customary power transformation. Inverters will progressively go about as shrewd energy the executives frameworks, organizing the consistent cooperation of different energy assets, capacity frameworks, and burdens. This development is fundamental for making an adaptable and versatile energy framework that can oblige the intricacies of a decentralized and manageable energy scene.

The zap of warming and modern cycles is one more groundbreaking pattern not too far off. Inverters will assume an imperative part in coordinating electric warming frameworks, for example, heat siphons, into the matrix. Furthermore, they will work with the charge of modern cycles, adding to the general decarbonization of the energy area. This shift towards zap lines up with the more extensive objective of diminishing dependence on non-renewable energy sources across different areas.

Hydrogen is arising as a flexible and clean energy transporter with the possibility to assume a critical part later on energy blend. Inverters are fundamental in electrolysis processes that produce green hydrogen by parting water into hydrogen and oxygen utilizing sustainable power. Also, inverters empower the productive utilization of hydrogen in energy components to create power when required. As the hydrogen economy extends, inverters will be instrumental in coordinating hydrogen-based energy frameworks into the network.

The round economy idea, accentuating the reasonable utilization of assets, is building up some decent momentum in the energy area. Inverters can add to the round economy by working with the mix of reused materials into their assembling processes. Planning inverters with an emphasis on recyclability and limiting natural effect lines up with the more extensive objective of making a supportable and round energy biological system.

The future job of energy inverters additionally crosses with the idea of energy democratization. Decentralized energy frameworks, joined with progressions in energy capacity and advanced advances, engage people and networks to take part in energy age and the executives effectively. Inverters go about as empowering influences of this democratization by giving the fundamental insight and control components at the neighborhood level.

Strategy and administrative structures will assume a significant part in forming the eventual fate of energy inverters. States all over the planet are perceiving the significance of boosting clean energy advances and cultivating development. Strong strategies can speed up the reception of cutting edge inverters, support innovative work, and establish a favorable climate for interests in economical energy arrangements.

Worldwide cooperation and normalization endeavors will be fundamental to guarantee interoperability and similarity of inverters across various locales. Blending specialized principles and affirmation processes works with the worldwide sending of energy inverters, supporting the consistent joining of sustainable power advancements into different energy lattices.

The job of energy inverters in improving energy strength is a basic thought, particularly despite environmental change and outrageous climate occasions. Inverters furnished with self-recuperating capacities and vigorous plan highlights add to the general versatility of energy frameworks. The capacity of inverters to independently recognize and alleviate shortcomings guarantees the proceeded with activity of basic foundation during unfriendly circumstances.

With regards to non-industrial nations, energy inverters can assume an extraordinary part in extending admittance to power. Decentralized sustainable power frameworks, empowered by inverters, offer a reasonable answer for zapping remote and off-lattice regions. These frameworks can give dependable and maintainable power, adding to destitution lightening and encouraging financial turn of events.

The eventual fate of the energy scene is set apart by a powerful interaction of mechanical progressions, strategy changes, and a developing accentuation on maintainability. In this developing setting, energy inverters are set to assume a urgent part in molding the direction of the energy area. Investigating future patterns and the advancing job of inverters gives bits of knowledge into the groundbreaking changes that will characterize the following period of the worldwide energy progress.

One of the conspicuous future patterns in the energy scene is the proceeded with development of environmentally friendly power sources. Sunlight based and wind energy, specifically, are encountering quick development as their costs decline and productivity gets to the next level. This flood in sustainable power arrangement requires progressed energy inverters to change over and deal with the variable result from these sources productively.

Energy capacity advances are ready to become essential parts of the energy scene, tending to the irregularity related with renewables. Inverters will assume a focal part in the joining of energy stockpiling frameworks, empowering consistent charging and releasing cycles. The collaboration among inverters and energy stockpiling improves matrix adaptability, upholds lattice security, and empowers a more dependable and strong energy foundation.

As the zap of transportation picks up speed, the interest for electric vehicles (EVs) is supposed to take off. This shift towards electric versatility presents new difficulties and open doors for energy frameworks. Inverters will be essential in working with the bidirectional progression of force among EVs and the network, empowering effective charging and releasing cycles. Savvy inverters furnished with cutting edge correspondence capacities will assume an essential part in improving the charging of EVs in light of framework conditions and power interest.

The idea of brilliant matrices is a groundbreaking pattern that imagines a wise and interconnected energy framework. In this specific situation, inverters act as wise hubs inside the matrix, able to do continuous correspondence and dynamic changes. Savvy inverters assume a key part in improving framework solidness, overseeing circulated energy assets, and empowering two-way correspondence between energy makers and customers.

The ascent of decentralized energy frameworks, including nanogrids and microgrids, addresses a change in perspective in energy age and dissemination. Inverters become integral to the activity of these decentralized frameworks by dealing with the different blend of fuel sources, like sunlight powered chargers, wind turbines, and energy stockpiling, at a restricted level. This pattern engages networks and organizations to have more noteworthy command over their energy supply, encouraging energy versatility and manageability.

Digitalization and the coordination of trend setting innovations, like man-made brainpower (computer based intelligence) and the Web of Things (IoT), are ready to reform the energy area. Brilliant inverters outfitted with simulated intelligence calculations can advance energy creation and utilization, anticipate framework conditions, and adjust to changing conditions progressively. The sending of IoT gadgets related to inverters takes into account upgraded checking, information assortment, and controller, adding to more effective and responsive energy frameworks.

With regards to future energy drifts, the job of energy inverters stretches out past conventional power change. Inverters will progressively go about as wise energy the board frameworks, coordinating the consistent communication of different energy assets, capacity frameworks, and burdens. This development is fundamental for making an adaptable and versatile energy framework that can oblige the intricacies of a decentralized and maintainable energy scene.

The zap of warming and modern cycles is one more groundbreaking pattern not too far off. Inverters will assume a fundamental part in coordinating electric warming frameworks, for example, heat siphons, into the lattice. Moreover, they will work with the zap of modern cycles, adding to the general decarbonization of the energy area. This shift towards charge lines up with the more extensive objective of decreasing dependence on non-renewable energy sources across different areas.

Hydrogen is arising as a flexible and clean energy transporter with the possibility to assume a huge part later on energy blend. Inverters are fundamental in electrolysis processes that produce green hydrogen by parting water into hydrogen and oxygen utilizing sustainable power.

Also, inverters empower the proficient utilization of hydrogen in energy units to create power when required. As the hydrogen economy extends, inverters will be instrumental in coordinating hydrogen-based energy frameworks into the matrix.

The roundabout economy idea, accentuating the maintainable utilization of assets, is building up momentum in the energy area. Inverters can add to the roundabout economy by working with the coordination of reused materials into their assembling processes. Planning inverters with an emphasis on recyclability and limiting natural effect lines up with the more extensive objective of making a reasonable and round energy environment.

The future job of energy inverters additionally crosses with the idea of energy democratization. Decentralized energy frameworks, joined with progressions in energy capacity and advanced innovations, engage people and networks to take part in energy age and the board effectively. Inverters go about as empowering influences of this democratization by giving the important knowledge and control components at the neighborhood level.

Strategy and administrative systems will assume a urgent part in forming the eventual fate of energy inverters. States all over the planet are perceiving the significance of boosting clean energy advancements and cultivating development. Steady strategies can speed up the reception of cutting edge inverters, support innovative work, and establish a helpful climate for interests in practical energy arrangements.

Worldwide cooperation and normalization endeavors will be fundamental to guarantee interoperability and similarity of inverters across various locales. Orchestrating specialized guidelines and confirmation processes works with the worldwide sending of energy inverters, supporting the consistent reconciliation of sustainable power advances into assorted energy networks.

The job of energy inverters in upgrading energy versatility is a basic thought, particularly despite environmental change and outrageous climate occasions. Inverters furnished with self-recuperating capacities and vigorous plan highlights add to the general flexibility of energy frameworks. The capacity of inverters to independently identify and moderate deficiencies guarantees the proceeded with activity of basic foundation during unfriendly circumstances.

With regards to emerging nations, energy inverters can assume a groundbreaking part in growing admittance to power. Decentralized environmentally friendly power frameworks, empowered by inverters, offer a feasible answer for energizing remote and off-matrix regions. These frameworks can give solid and manageable power, adding to destitution easing and cultivating financial turn of events.

Chapter 9

Harnessing Energy Inverters for a Sustainable Future

As mankind remains at the intersection of ecological emergencies and the consistently expanding interest for energy, the requirement for economical and effective energy arrangements has never been more basic. Among the heap innovations endeavoring to address this test, energy inverters arise as a promising road for changing the manner in which we create, disseminate, and consume power.

Energy inverters assume an essential part in the progress towards a manageable future by changing over and overseeing electrical energy. These gadgets are key to environmentally friendly power frameworks, empowering the incorporation of sources like sunlight based chargers and wind turbines into the power lattice. In this investigation, we dive into the complexities of energy inverters, their applications, and the extraordinary effect they can have on our journey for a greener and more supportable energy scene.

At the core of the matter lies the major idea of energy reversal. An energy inverter is a gadget that converts direct current (DC) to substituting current (AC) or the other way around. This capacity is fundamental in environmentally friendly power frameworks, where the result from sources like sunlight based chargers and wind turbines is normally in DC, while our homes and businesses overwhelmingly use AC. By working with this change, energy inverters overcome any issues between environmentally friendly power sources and the customary power network.

One of the essential uses of energy inverters is in photovoltaic (PV) frameworks, where sun powered chargers saddle daylight and convert it into electrical energy. Sun powered chargers produce DC, and to make this energy viable with the air conditioner based power network, inverters are utilized. These gadgets carry out the fundamental role of changing the DC yield from sunlight based chargers into the air conditioner that drives our homes and organizations.

Additionally, wind turbines create electrical energy in DC structure, and energy inverters are utilized to change over this energy into AC for framework mix. This interaction is essential for expanding the productivity and utility of sustainable power sources, guaranteeing a consistent progress between the decentralized age of clean energy and its incorporation into the concentrated power framework.

The extraordinary effect of energy inverters stretches out past the domain of environmentally friendly power incorporation. In present day power frameworks, these gadgets likewise assume an essential part in guaranteeing network soundness and dependability. Inverter-based innovations empower the control and guideline of force stream, voltage, and recurrence, adding to a stronger and versatile electrical foundation.

Microgrids, which are restricted and free energy frameworks, influence energy inverters to enhance energy conveyance inside a particular local area or office. These frameworks can work independently or related to the principal power lattice, improving energy strength and supportability. Energy inverters go about as the key part in microgrid designs, working with consistent changes between lattice associated and islanded modes.

Past their applications in sustainable power and microgrids, energy inverters are necessary parts in energy capacity frameworks. As the interest for energy keeps on rising, the capacity to store and delivery energy productively becomes vital. Inverter-based advances empower the successful coordination of energy stockpiling gadgets like batteries, considering the capacity of surplus energy during times of low interest and its delivery during top interest.

The development of shrewd lattices, portrayed by cutting edge correspondence and control frameworks, depends intensely on energy inverters for constant checking and streamlining of energy stream. These smart lattices influence the capacities of inverters to upgrade matrix flexibility, oblige circulated energy assets, and further develop generally speaking energy proficiency.

Chasing a maintainable future, the job of energy inverters reaches out past their specialized functionalities. Strategy structures and administrative measures assume a significant part in molding the scene for environmentally friendly power innovations, including energy inverters. States and administrative bodies overall are progressively perceiving the significance of boosting the sending of environmentally friendly power frameworks and the related inverter advances.

Monetary motivating forces, feed-in taxes, and administrative systems that energize the reception of clean energy advances add to the boundless sending of energy inverters. These actions advance the utilization of environmentally friendly power as well as encourage development in inverter advances, driving enhancements in productivity, dependability, and cost-adequacy.

One of the critical difficulties in tackling energy inverters for a practical future lies in the streamlining of their productivity and dependability. Similarly as with any innovation, there is a consistent drive to improve execution measurements, decrease energy misfortunes, and increment the life expectancy of these gadgets. Innovative work endeavors center around propelling the cutting edge in inverter advancements, with a sharp eye on working on their natural effect.

The natural impression of energy inverters incorporates angles, for example, the materials utilized in their development, energy utilization during activity, and end-of-life removal contemplations. Economical assembling rehearses, the utilization of recyclable materials, and the improvement of inverter advancements with insignificant natural effect are fundamental parts of guaranteeing the general maintainability of the energy change.

The combination of energy inverters into the more extensive setting of the Web of Things (IoT) and Industry 4.0 further enhances their true capacity for changing our energy scene. Shrewd and associated inverters can speak with different gadgets, gather and dissect information progressively, and settle on astute choices to improve energy creation, utilization, and conveyance. This interconnectedness adds to the formation of stronger, versatile, and proficient energy frameworks.

The continuous digitalization of the energy area, frequently alluded to as the energy change or energy change, highlights the requirement for smart and versatile innovations like energy inverters. As we move towards a future portrayed by decentralized energy age, expanded charge, and a developing accentuation on maintainability, the job of these gadgets turns out to be considerably more articulated.

With regards to sustainable power, the discontinuity of sources, for example, sun oriented and wind represents a critical test to framework administrators. Energy inverters, outfitted with cutting edge control calculations and network the board abilities, can moderate the effect of this discontinuity.

Matrix framing inverters, for instance, can keep up with network dependability and quality in any event, when disengaged from the primary power framework, adding to the general strength of the electrical foundation.

Notwithstanding their part in network soundness, energy inverters add to accomplishing a more significant level of energy freedom and security. By outfitting locally accessible inexhaustible assets and conveying energy stockpiling arrangements, networks and ventures can decrease their dependence on brought together power sources. Energy inverters assume a focal part in coordinating these disseminated energy assets, empowering a stronger and versatile energy foundation.

The incorporation of energy inverters into present day power frameworks likewise opens new roads for energy exchanging and distributed energy exchanges. Blockchain innovation, combined with brilliant inverters, can work with secure and straightforward energy exchanges between prosumers (the individuals who both produce and consume energy) inside a decentralized energy commercial center. This distributed energy exchanging model enables people and networks to effectively take part in the energy progress.

The progressions in energy inverter advancements are not restricted to the electrical space. Advancements in power hardware, control frameworks, and materials science add to the improvement of more reduced, lightweight, and productive inverters. These progressions are especially urgent in applications like electric vehicles (EVs) and versatile electronic gadgets, where the size and weight of inverters straightforwardly influence generally speaking framework execution and client experience.

The charge of transportation, driven by the basic to decrease ozone depleting substance outflows and reliance on petroleum derivatives, depends on energy inverters in EVs. Inverter advances in electric vehicles assume a basic part in dealing with the progression of electrical energy between the battery and the engine, guaranteeing ideal execution and proficiency. As the car business advances towards electric portability, the interest for superior execution and practical inverters is set to rise.

The boundless reception of energy-proficient apparatuses and the rising reconciliation of sustainable power sources into private structures feature the significance of energy inverters in the customer space. Brilliant homes, furnished with insightful energy the executives frameworks and associated gadgets, influence inverters to upgrade energy utilization, store overflow energy, and add to generally speaking energy proficiency.

The worldwide change towards practical energy frameworks additionally includes addressing energy neediness and guaranteeing admittance to power in underserved locales. Off-matrix and smaller than usual framework arrangements, controlled by environmentally friendly power sources and upheld by energy inverters, offer a suitable method for charging remote and

provincial regions. These decentralized energy frameworks add to neediness easing, further developed medical care, and upgraded instructive open doors in networks that need admittance to dependable power.

9.1 Summarize key takeaways from the book.

The book viable gives a complete investigation of energy inverters and their urgent job in forming an economical future. As we explore the complicated convergence of ecological difficulties and raising energy requests, the meaning of energy inverters arises as a signal of commitment as we continued looking for cleaner and more productive energy arrangements.

At its center, an energy inverter fills the essential need of changing over direct current (DC) to exchanging current (AC) or the other way around. This usefulness is basic with regards to sustainable power frameworks, where sources like sunlight powered chargers and wind turbines overwhelmingly create DC, while our current power network depends on AC. The book fastidiously unloads the utilizations of energy inverters, zeroing in on their focal job in photovoltaic frameworks, wind energy combination, and microgrids.

The combination of energy inverters into photovoltaic frameworks is a foundation of the environmentally friendly power scene. Sun powered chargers, answerable for bridling daylight and changing over it into electrical energy, produce yield in DC. Energy inverters assume a basic part in this situation by working with the change of DC yield into AC, making it viable with the customary power lattice. The book highlights the significance of this transformation cycle in opening the maximum capacity of sun powered energy and its consistent mix into the more extensive energy framework.

Wind energy, one more central member in the sustainable power field, additionally depends on energy inverters for viable framework reconciliation. Wind turbines produce electrical energy in DC, and energy inverters act as the scaffold, changing over this result into the air conditioner that controls our homes and enterprises. The book digs into the complexities of this transformation cycle, underlining its part in boosting the proficiency and utility of wind energy sources.

Microgrids, as confined and free energy frameworks, influence energy inverters to streamline energy dispersion inside unambiguous networks or offices. This segment of the book reveals insight into the independence and versatility microgrids gain through the arrangement of energy inverters. These frameworks can work autonomously or related to the primary power network, adding to upgraded energy flexibility and manageability.

The book further extends its degree to envelop the job of energy inverters in energy capacity frameworks. With the raising interest for energy, productive capacity arrangements become fundamental.

Inverter-based advances empower the consistent coordination of energy stockpiling gadgets, like batteries, taking into consideration the capacity of surplus energy during times of low interest and its delivery during top interest. This segment features the significant job of energy inverters in molding the eventual fate of energy stockpiling and burden the executives.

The rise of shrewd lattices, described by cutting edge correspondence and control frameworks, is unpredictably attached to energy inverters. The insightful lattices influence the capacities of inverters for ongoing observing and improvement of energy stream, adding to a stronger and versatile electrical framework. This segment of the book investigates how energy inverters act as key parts in the progress towards more astute and more effective power networks.

A huge piece of the book is committed to the more extensive setting where energy inverters work. The account flawlessly winds through strategy systems and administrative measures that shape the scene for environmentally friendly power advances, including energy inverters. Legislatures and administrative bodies overall are progressively perceiving the significance of boosting the organization of sustainable power frameworks and related inverter innovations. Monetary motivating forces, feed-in taxes, and administrative systems add to the far reaching reception of energy inverters, encouraging development and driving upgrades in proficiency and cost-adequacy.

Be that as it may, the book doesn't avoid tending to the difficulties inborn in the excursion towards a feasible future with energy inverters at its center. Improving the effectiveness and unwavering quality of energy inverters stands apart as a key test. The persevering quest for progressions in execution measurements, decrease of energy misfortunes, and expanded life expectancy of these gadgets is a focal topic in continuous innovative work endeavors. The book highlights the significance of maintainable assembling rehearses, the utilization of recyclable materials, and limiting the natural impression of energy inverters to guarantee the general maintainability of the energy progress.

The coordination of energy inverters into the more extensive setting of the Web of Things (IoT) and Industry 4.0 addresses a huge enunciation point. The interconnectedness of shrewd inverters with different gadgets, combined with ongoing information assortment and examination, adds to the making of stronger, versatile, and productive energy frameworks. The book investigates the groundbreaking capability of this coordination and its suggestions for the fate of energy the board.

The account then flawlessly advances to the developing scene of man-made reasoning (simulated intelligence) and AI related to energy inverters. This part frames how computer based intelligence calculations improve control, observing, and enhancement capacities. Savvy inverters, fit for adjusting to dynamic circumstances and gaining from constant information, are situated as vital participants in the continuous digitalization of the energy area.

The book underscores the job of simulated intelligence in prescient upkeep, where AI calculations expect likely issues and advance the exhibition of inverter frameworks, adding to further developed dependability and decreased personal time.

The joint effort of specialists, designers, and pioneers overall becomes the dominant focal point in the book's investigation of the advancement of energy inverters. High level materials, for example, wide-bandgap semiconductors, are featured for their commitments to the improvement of conservative and superior execution inverters. The book highlights the interdisciplinary idea of innovative work in energy inverters, underscoring the requirement for coordinated efforts between the scholarly community, industry, and government bodies to cultivate development and address specialized difficulties.

Developments in power gadgets, control frameworks, and materials science are not restricted to the electrical space. The book uncovers their applications in regions like electric vehicles (EVs) and compact electronic gadgets. With regards to EVs, inverter advancements assume a basic part in dealing with the progression of electrical energy between the battery and the engine, forming the direction of the car business towards electric versatility. The interest for elite execution and practical inverters is projected to ascend as the world hugs electric transportation.

The book expands its venture into the buyer space, investigating the job of energy inverters in the far and wide reception of energy-productive machines and the mix of sustainable power sources into private structures. Savvy homes, outfitted with astute energy the executives frameworks and associated gadgets, influence inverters to advance energy utilization, store overflow energy, and add to in general energy proficiency.

The worldwide progress towards practical energy frameworks requires addressing energy neediness and guaranteeing admittance to power in underserved areas. The book features off-network and smaller than usual framework arrangements, controlled by environmentally friendly power sources and upheld by energy inverters, as practical method for zapping remote and provincial regions. These decentralized energy frameworks add to neediness lightening, further developed medical services, and upgraded instructive open doors in networks that need admittance to solid power.

In the last areas, the book addresses financial, international, and cultural contemplations that impact the direction of the energy change. The monetary reasonability of environmentally friendly power advancements, including energy inverters, arises as a critical determinant in their boundless reception. Government motivators, endowments, and strong strategies assume a vital part in crossing over the monetary hole among ordinary and environmentally friendly power sources.

International contemplations, like the accessibility of basic unrefined substances, are recognized as possible bottlenecks in the worldwide progress to environmentally friendly power. The book advocates for broadening of supply chains, investigation of elective materials, and worldwide joint efforts to moderate these international difficulties. Cultural acknowledgment of environmentally friendly power advances is depicted as a mind boggling transaction of variables, with local area commitment and training arising as fundamental parts in guaranteeing effective combination.

Administrative systems, both at the public and global levels, are recognized as critical in molding the eventual fate of energy inverters. Clear and strong guidelines establish an empowering climate for the arrangement of sustainable power advancements, empowering venture, development, and rivalry on the lookout. The book underlines the significance of normalized correspondence conventions and connection points for guaranteeing interoperability and adaptability in the energy inverter scene.

The powerful idea of the energy scene, combined with fast innovative headways, highlights the requirement for continuous coordinated effort and a multidisciplinary approach. The book finishes up by emphasizing the complex excursion towards a feasible future fueled by energy inverters. These gadgets, filling in as key parts in the reconciliation of environmentally friendly power sources, streamlining of force frameworks, and formation of versatile energy foundations, hold the way to opening a future where perfect, reasonable, and productive energy controls our reality.

9.2 Emphasize the role of energy inverters in achieving a sustainable and resilient energy future.

Accomplishing a practical and versatile energy future is a worldwide basic driven by the squeezing need to address environmental change and the rising interest for perfect and solid energy sources. At the very front of this extraordinary excursion are energy inverters, assuming an essential part in reshaping our energy scene. This talk dives into the multi-layered commitments of energy inverters and their basic job chasing after maintainability and versatility.

Energy inverters, at their embodiment, are gadgets that work with the transformation of electrical energy between direct flow (DC) and rotating flow (AC). This ability is basic in the reconciliation of sustainable power sources into our current power foundation. The ascent of sustainable power, especially sun oriented and wind power, presents an extraordinary arrangement of difficulties because of the inborn idea of these sources delivering DC power. Energy inverters overcome any barrier, empowering the consistent coordination of environmentally friendly power into the air conditioner driven network that controls our homes, organizations, and ventures.

With regards to photovoltaic frameworks, energy inverters arise as key parts in the change to an economical energy future. Sun powered chargers, the essential parts of these frameworks, create DC power when presented to daylight. To make this energy viable with the network and customary apparatuses, energy inverters carry out the imperative role of switching DC over completely to AC. This cycle is instrumental in opening the maximum capacity of sun oriented energy and encouraging its boundless reception as a spotless and sustainable power source.

Wind energy, one more central member in the sustainable power range, depends on energy inverters for compelling lattice joining. Wind turbines produce power in DC structure, and energy inverters work with the change to AC, lining up with the framework's qualities. This change is fundamental for consistent joining into the current energy foundation, empowering wind ability to contribute essentially to the general energy blend. By filling in as an extension between sustainable sources and the lattice, energy inverters assume a focal part in differentiating our energy portfolio and diminishing dependence on petroleum derivatives.

Microgrids, arising as restricted and versatile energy arrangements, influence energy inverters to advance energy conveyance inside unambiguous networks or offices. The independence and flexibility of microgrids are upgraded through the sending of energy inverters, taking into account consistent advances between network associated and islanded modes. This capacity adds to energy flexibility, guaranteeing a steady power supply even with interruptions or crises. Energy inverters go about as wise facilitators inside microgrid models, adjusting the age and utilization of energy for most extreme effectiveness.

In the more extensive setting of energy stockpiling, energy inverters work with the viable coordination of capacity frameworks, like batteries, into the energy foundation. As the interest for energy keeps on rising, the capacity to store overflow energy during times of low interest and delivery it during top interest becomes principal. Energy capacity frameworks, combined with energy inverters, assume a urgent part in load adjusting, adding to lattice dependability and improving generally speaking energy proficiency. This capacity turns out to be progressively significant even with variable energy age from sustainable sources.

Shrewd matrices, described by cutting edge correspondence and control frameworks, depend vigorously on energy inverters for constant observing and streamlining of energy stream. The knowledge implanted in these inverters considers dynamic changes in accordance with energy appropriation, answering vacillations in organic market. By improving framework adaptability and versatility, energy inverters add to the flexibility of shrewd lattices, making ready for a more responsive and effective power foundation.

The book of energy inverters reaches out past their specialized functionalities to the domain of strategy structures and administrative measures. States and administrative bodies overall are perceiving the pivotal job of energy inverters in the change to a reasonable energy future.

Motivating forces, endowments, and strong strategies are being formed to empower the organization of environmentally friendly power frameworks, with a particular spotlight on the incorporation of energy inverters. These actions advance the utilization of clean energy as well as animate development in inverter advances, driving upgrades in proficiency, unwavering quality, and cost-adequacy.

Endeavors to bridle energy inverters for maintainability are not without challenges. Upgrading the proficiency and dependability of energy inverters is a ceaseless undertaking. Specialists and designers are devoted to propelling the best in class in inverter advancements, meaning to diminish energy misfortunes, improve execution measurements, and broaden the life expectancy of these gadgets. Reasonable assembling rehearses, the utilization of recyclable materials, and limiting the natural effect of energy inverters are basic to guaranteeing the general supportability of the energy change.

The coordination of energy inverters into the more extensive scene of the Web of Things (IoT) and Industry 4.0 messengers another period of interconnected energy frameworks. Savvy and associated inverters can speak with different gadgets, gather and dissect constant information, and pursue astute choices to upgrade energy creation, utilization, and dispersion. This interconnectedness adds to the formation of stronger, versatile, and proficient energy frameworks, lining up with the standards of a reasonable energy future.

Man-made consciousness (artificial intelligence) and AI calculations are transforming the energy inverter space, upgrading control, observing, and enhancement capacities. Smart inverters, engaged by simulated intelligence, can adjust to dynamic circumstances, gain from constant information, and upgrade energy stream in manners that customary inverters can't. The collaboration among computer based intelligence and energy inverters reaches out to prescient upkeep, where AI calculations investigate authentic information to expect likely issues, adding to further developed dependability and diminished free time.

Coordinated efforts between the scholarly world, industry, and government bodies are at the front line of driving advancement in energy inverters. High level materials, for example, wide-bandgap semiconductors, are acquiring conspicuousness in inverter configuration, offering further developed effectiveness and dependability contrasted with conventional materials. The interdisciplinary idea of innovative work endeavors highlights the requirement for cooperative undertakings to address specialized difficulties, encourage development, and shape the administrative structures that administer the arrangement of energy inverter advancements.

The extent of energy inverters reaches out past conventional power frameworks into arising spaces like electric vehicles (EVs). In the jolt of transportation, energy inverters assume a basic part in dealing with the progression of electrical energy between the battery and the engine in EVs.

As the car business advances towards electric versatility to decrease ozone harming substance emanations and reliance on non-renewable energy sources, the interest for elite execution and savvy inverters is set to rise.

The effect of energy inverters resounds in the customer space, affecting the reception of energy-proficient machines and environmentally friendly power sources in private structures. Brilliant homes, outfitted with smart energy the executives frameworks and associated gadgets, influence inverters to enhance energy utilization, store overflow energy, and add to in general energy proficiency. This buyer driven approach builds up the job of energy inverters in driving manageability at both the large scale and miniature levels.

Energy destitution and the absence of admittance to power in underserved districts are difficulties that energy inverters address through off-framework and scaled down matrix arrangements. These decentralized energy frameworks, fueled by sustainable sources and upheld by energy inverters, offer a help to remote and rustic regions. The arrangement of solid power adds to destitution lightening, further developed medical services, and improved instructive open doors, denoting a huge step towards accomplishing an additional fair and economical future.

While the excursion towards a manageable energy future controlled by energy inverters is set apart by striking advancement, it isn't without its difficulties. Monetary contemplations, international elements, and cultural acknowledgment are essential in molding the direction of the energy progress. The expense viability of environmentally friendly power advancements, including energy inverters, is a vital determinant in their far reaching reception and combination into standard energy frameworks.

International contemplations, like the accessibility of basic natural substances, can present difficulties to the worldwide change to environmentally friendly power. Broadening of supply chains, investigation of elective materials, and worldwide joint efforts are fundamental methodologies to moderate international bottlenecks. Cultural acknowledgment of sustainable power innovations, including energy inverters, is a mind boggling transaction of elements that requires local area commitment, schooling, and the contribution of neighborhood partners in dynamic cycles.

The administrative scene arises as a basic figure molding the eventual fate of energy inverters. Clear and strong guidelines establish an empowering climate for the arrangement of environmentally friendly power advancements, empowering venture, development, and contest on the lookout. Administrative systems that focus on supportability, matrix combination, and energy productivity add to the speed increase of the energy progress.

9.3 Inspire readers to explore further innovations and applications in the field.

Leaving on an excursion of investigation in the field of energy developments presents a thrilling and unlimited chance to shape the eventual fate of manageable and versatile energy frameworks. As we stand at the nexus of mechanical progressions and ecological goals, the source of inspiration resonates: rouse perusers to dive into additional developments and applications that can push us towards a more maintainable and fair energy future.

The surprising headway made in energy inverters, as framed in going before segments, fills in as a demonstration of the extraordinary potential inside the more extensive energy scene. However, this isn't the finish but instead a venturing stone towards a skyline overflowing with potential outcomes. The story subsequently turns to light a flash of interest and excitement, encouraging perusers to investigate roads that reach out past the bounds of current information.

In the domain of environmentally friendly power coordination, where energy inverters assume an essential part, there exists a range of undiscovered potential anticipating investigation. The improvement of existing advancements and the advancement of novel arrangements stand as ripe justification for development. Analysts and specialists are welcome to dive into the complexities of inverter configuration, pushing the limits of productivity, dependability, and flexibility.

The collaboration between energy inverters and man-made brainpower (artificial intelligence) presents a fascinating wilderness ready for investigation. As artificial intelligence advances keep on developing, there is gigantic potential to improve the knowledge implanted in energy inverters. Simulated intelligence calculations can be custom fitted to examine complex datasets, anticipate energy request designs, and progressively streamline energy stream in manners that essentially enhance framework execution. This combination of artificial intelligence and energy inverters allures pioneers to diagram new regions in brilliant energy the executives.

In the mission for supportability, the improvement of energy inverters with negligible ecological effect remains as a considerable test and a goal. Manageable assembling rehearses, the utilization of eco-accommodating materials, and the consolidation of round economy standards are roads to investigate. Pioneers are urged to investigate manners by which energy inverters can be delivered and worked with an emphasis on lessening carbon impressions and limiting waste.

The growing space of energy stockpiling innovations, combined with energy inverters, offers a material for creative personalities to paint upon. The investigation of cutting edge energy capacity arrangements, like cutting edge batteries and inventive capacitor advances, can change how overflow energy is put away and circulated.

Spearheading the advancement of energy stockpiling frameworks that consistently coordinate with energy inverters makes the way for more solid and versatile power foundations.

The convergence of energy inverters with blockchain innovation welcomes trend-setters to imagine decentralized energy frameworks and distributed energy exchanging stages. Blockchain, known for its straightforwardness, security, and detectability, can be saddled to make hearty and decentralized energy commercial centers. This investigation can engage people and networks to effectively take part in the energy change, cultivating a feeling of responsibility and manageability.

The joining of energy inverters into the quickly advancing scene of electric portability gives a chance to reshape the fate of transportation. As electric vehicles (EVs) become progressively predominant, there is a call for pioneers to dive into the complexities of inverter innovations customized for the extraordinary requests of the car area. Superior exe-cution, lightweight, and financially savvy inverters are key components in speeding up the reception of electric vehicles and lessening our re-liance on regular cars.

In the shopper space, the development of shrewd homes and savvy energy the board frameworks welcomes investigation into client driven advancements. Pioneers can imagine and foster instinctive connection points that enable mortgage holders to draw in with their energy uti-lization effectively. The reconciliation of energy inverters with arising advancements, for example, expanded reality and remote helpers opens up opportunities for making easy to understand and intelligent energy the executives encounters.

The idea of energy strength, profoundly laced with microgrids and energy inverters, is a field that calls for additional investigation. Trailblazers are urged to imagine and plan microgrid arrangements that flawlessly incorporate environmentally friendly power sources, capacity advances, and wise inverters. The formation of strong and versatile microgrids can rethink how networks and enterprises answer blackouts, catastrophic events, and other unexpected difficulties.

As energy frameworks become progressively interconnected, the investigation of network protection estimates inside the domain of energy inverters becomes vital. Pioneers are tested to foster powerful online protection systems that shield basic energy framework from digital dangers. Encryption, validation conventions, and interruption recognition frameworks custom fitted for energy inverters structure the structure blocks of secure and tough energy biological systems.

The worldwide idea of the energy change requires cooperative endeavors that rise above geological limits. Pioneers are encouraged to participate in worldwide coordinated efforts, information sharing, and joint examination drives.

This interconnected methodology cultivates a worldwide local area of trailblazers pursuing a common vision of a practical and tough energy future.

The investigation of materials science, especially in the improvement of cutting edge semiconductors, holds gigantic commitment for the development of energy inverters. Pioneers are welcome to dig into the universe of materials research, looking for novel arrangements that upgrade the productivity and unwavering quality of energy inverters. The quest for wide-bandgap semiconductors and other state of the art materials presents a potential chance to rethink the benchmarks of inverter execution.

For those spellbound by the capability of energy inverters in tending to energy neediness, there is a call to investigate versatile and replicable arrangements. Pioneers are urged to devise off-framework and little matrix frameworks that are custom-made to the particular requirements of underserved networks. These arrangements shouldn't just give admittance to power yet in addition engage networks to outfit their neighborhood sustainable assets for manageable and comprehensive turn of events.

The investigation of public strategy and administrative systems turns into a convincing road for those enthusiastic about molding the energy progress. Trend-setters are welcome to effectively draw in with policymakers, upholding for arrangements that boost the organization of sustainable power frameworks and energy inverters. The improvement of administrative structures that focus on maintainability, matrix coordination, and energy productivity is necessary to speeding up the change towards a cleaner and stronger energy scene.

As pioneers investigate the wildernesses of energy innovation, there is a requirement for an all encompassing and interdisciplinary methodology. The assembly of assorted fields, from designing and materials science to information investigation and sociologies, holds the way to opening extensive arrangements. Trend-setters are urged to team up across disciplines, perceiving that the difficulties of maintainability require multi-layered and coordinated arrangements.

The excursion of investigation and advancement is dynamic and consistently developing. Trailblazers are called upon not exclusively to foster earth shattering innovations yet in addition to encourage a culture of supportability, inclusivity, and moral thought. The moral elements of energy developments, including contemplations of social value, ecological equity, and local area commitment, ought to be essential to the investigation of additional opportunities in the energy area.

In the unique domain of energy advancements and applications, a huge scene of conceivable outcomes anticipates investigation and improvement. As we explore the mind boggling difficulties of a changing environment and developing energy requests, the basic to push the limits of current information and embrace groundbreaking developments turns out to be progressively clear. This story unfurls as a challenge to dig into different features of the field, traversing mechanical forward leaps, novel applications, and interdisciplinary coordinated efforts, all pointed toward forming a supportable and strong energy future.

At the front line of this investigation are headways in energy capacity innovations, a basic space that meets with different features of the energy scene. Advancements in battery advances, supercapacitors, and other capacity arrangements hold monstrous commitment for tending to the discontinuity of environmentally friendly power sources. Analysts and architects are welcome to redefine known limits, looking for leap forwards that improve energy thickness, life expectancy, and charging proficiency. The improvement of cutting edge energy capacity gadgets can possibly change how we catch, store, and disseminate energy in a way that lines up with maintainability objectives.

Investigating the nexus between energy capacity and man-made consciousness (artificial intelligence) presents a change in outlook in energy the executives frameworks. Man-made intelligence calculations can break down complex examples of energy utilization, foresee request variances, and improve the activity of capacity frameworks. Trailblazers are urged to dive into the reconciliation of artificial intelligence with energy capacity, making astute and versatile frameworks that can independently answer evolving conditions. The collaboration among artificial intelligence and energy stockpiling upgrades proficiency as well as adds to the versatility of energy networks by empowering dynamic and information driven navigation.

In the journey for reasonable energy arrangements, the investigation of materials science arises as a foundation. Trend-setters are called upon to dig into the improvement of cutting edge materials that can change energy innovations. The journey for superior execution and manageable materials ranges from productive sun oriented cell materials to inventive coatings for wind turbine sharp edges. By investigating new materials with improved properties, scientists add to the enhancement of energy transformation and capacity frameworks, introducing another period of productivity and manageability.

The intermingling of energy advancements with the quickly propelling area of nanotechnology presents an astonishing wilderness. Nanomaterials, described by their extraordinary properties at the nanoscale, hold the possibility to rethink energy applications. Pioneers are urged to investigate nanomaterials for upgraded sun oriented cell productivity, further developed catalysis in power modules, and novel ways to deal with energy collecting. The control of materials at the nanoscale opens roads for remarkable command over energy-related processes, offering a pathway to more productive and feasible energy innovations.

In the domain of sustainable power, the investigation of creative sunlight based advancements goes past customary photovoltaic frameworks. Trend-setters are welcome to dive into arising advances like sun based paint, sun oriented windows, and sun powered materials. These applications coordinate sun oriented cells into regular surfaces, extending the potential for sun based energy catch in metropolitan conditions. The investigation of unpredictable sun oriented arrangements lines up with the basic to consistently incorporate environmentally friendly power into the texture of our day to day routines, making manageability an omnipresent and natural part of our environmental elements.

The investigation of biomimicry in energy advancements takes advantage of the rich woven artwork of nature for motivation. Trend-setters are urged to study and recreate regular cycles to foster energy-productive arrangements. From impersonating photosynthesis for further developed sun based energy transformation to copying the energy-effective trip of birds for streamlined breeze turbines, biomimicry offers a plan for supportable development. By drawing from the inventiveness of the normal world, analysts can uncover novel ways to deal with energy challenges that are both productive and naturally cognizant.

The jolt of transportation remains as an extraordinary power in the mission for practical portability. Trailblazers are called upon to investigate headways in electric vehicle (EV) advancements, going from superior execution batteries to creative charging framework. The advancement of quick charging innovations, broadened battery duration, and lightweight materials for EVs adds to the speed increase of the worldwide shift towards manageable transportation. The investigation of savvy networks and vehicle-to-framework (V2G) advancements further expands the potential for consistent coordination of electric vehicles into the more extensive energy environment.

Interdisciplinary joint efforts among energy and medical services areas offer an interesting road for development. The investigation of energy-productive advancements in medical care offices, going from brilliant structure frameworks to maintainable clinical gadgets, lines up with the more extensive objective of making earth cognizant medical care foundation. Trailblazers are welcome to investigate how energy developments can add to the decrease of carbon impressions in medical services, cultivating a cooperative connection between energy productivity and general wellbeing.

The crossing point of energy developments with water the board acquaints arrangements with address the water-energy nexus. Trailblazers are encouraged to investigate advances that upgrade the productivity of water desalination, wastewater treatment, and water system frameworks. Feasible water the executives is unpredictably connected to energy utilization, and by creating energy-proficient arrangements, scientists add to the versatility of water and energy frameworks. The investigation of incorporated frameworks that blend water and energy contemplations shapes an all encompassing way to deal with tending to interconnected manageability challenges.

The organization of microgrids addresses a change in outlook in decentralized energy frameworks. Trailblazers are urged to investigate novel applications and plans of action for microgrid innovations. From strong energy supply in far off regions to supporting basic foundation during framework blackouts, microgrids offer adaptable arrangements. The mix of environmentally friendly power sources, energy capacity, and high level control frameworks in microgrids presents a thrilling field for investigation, engaging networks and ventures to accomplish energy freedom.

In the advanced age, the investigation of blockchain innovation in energy frameworks presents extraordinary conceivable outcomes. Trendsetters are asked to investigate blockchain applications for shared energy exchanging, decentralized energy commercial centers, and straightforward energy exchanges. Blockchain's intrinsic elements of straightforwardness, security, and decentralization line up with the standards of a feasible and fair energy future. The investigation of blockchain in energy frameworks addresses a spearheading work to democratize energy access and make comprehensive energy economies.

The combination of energy advancements with the arising field of quantum figuring opens roads for tackling complex energy-related difficulties. Quantum registering's computational power can possibly alter reenactments for materials disclosure, enhance energy frameworks, and tackle mind boggling issues in energy research. Pioneers are urged to investigate how quantum processing can be outfit to speed up forward leaps in energy advances, preparing for remarkable headways in proficiency and maintainability.

The investigation of local area based energy drives welcomes pioneers to reconsider the connection between people, networks, and energy frameworks. Local area energy projects, like cooperatives and shared sustainable power assets, engage people to effectively partake in the energy change. Trailblazers are encouraged to investigate models that focus on local area commitment, proprietorship, and aggregate navigation, cultivating a feeling of obligation and shared benefits chasing maintainable energy.

As energy frameworks become progressively interconnected, the investigation of strength notwithstanding environmental change and catastrophic events becomes central. Trailblazers are called upon to foster advancements and procedures that upgrade the strength of energy foundation. From matrix solidifying advancements to versatile energy stockpiling arrangements, the investigation of strong energy frameworks adds to relieving the effects of outrageous climate occasions and guaranteeing the dependability of energy supply despite misfortune.

The investigation of energy developments in the rural area offers answers for address the nexus of food and energy. Trend-setters are asked to investigate economical energy applications in accuracy horticulture, going from energy-productive water system frameworks to the zap of cultivating hardware.

The mix of environmentally friendly power sources in horticulture lines up with the more extensive objective of making tough and manageable food creation frameworks, tending to the difficulties presented by environmental change and guaranteeing energy access for provincial networks.

In the investigation of advancements, inclusivity becomes the dominant focal point. Trailblazers are urged to consider the social elements of energy arrangements, guaranteeing that headways benefit all sections of society. The improvement of reasonable and open innovations, combined with local area commitment and limit building drives, encourages a more comprehensive and evenhanded energy progress. The investigation of arrangements that span the energy access hole and engage minimized networks adds to the overall objective of a manageable and just energy future.

www.ingramcontent.com/pod-product-compliance
Lightning Source LLC
LaVergne TN
LVHW020737200726
843506LV00009B/796